AF333237

Military Education

Military Education
A Reference Handbook

Cynthia A. Watson

Contemporary Military, Strategic, and Security Issues

PRAEGER SECURITY INTERNATIONAL
Westport, Connecticut • London

Library of Congress Cataloging-in-Publication Data

Watson, Cynthia Ann.
Military education : a reference handbook / Cynthia A. Watson.
     p. cm. – (Contemporary military, strategic, and security issues, ISSN 1932–295X)
    Includes bibliographical references and index.
    ISBN 0–275–99219–5 (alk. paper)
    1. Military education—United States—Handbooks, manuals, etc.   I. Title.
   U408.W37   2007
   355.0071'173–dc22   2006037687

British Library Cataloguing in Publication Data is available.

Library of Congress Catalog Card Number: 2006037687
ISBN-10: 0–275–99219–5
ISBN-13: 978–0–275–99219–4
ISSN: 1932–295X

First published in 2007

Praeger Security International, 88 Post Road West, Westport, CT 06881
An imprint of Greenwood Publishing Group, Inc.
www.praeger.com

Printed in the United States of America

The paper used in this book complies with the
Permanent Paper Standard issued by the National
Information Standards Organization (Z39.48–1984).

10 9 8 7 6 5 4 3 2 1

To the faculty of the National War College—
Educators and Students all

Contents

Preface

Ms. Alicia Merritt of Praeger/Greenwood Publishing was open to topics, which many editors would have dismissed. Her support has been invaluable and I thoroughly enjoy working with her. Dr. Sharon Murphy of Nazareth College and Dr. Paul H.B. Godwin are patient, tremendous colleagues who invariably have time to be friends. Scott and Bonnie Nordstrom are such wonderful supporters.

The National Defense University is a marvelous, diverse organization. I have had the opportunity to interact with many in various corners of the University and appreciate all I have learned from all of them. I consider many valued colleagues. A special note of thanks to the NDU Library and its staff. Dr. Sarah Mikel has assembled the best group of librarians anywhere in the world. Regardless of the subject, these individuals enthusiastically work with the faculty and students to make sure they have the opportunity to think instead of getting frustrated over not being able to find a book or article.

My colleague Dr. John Yaeger, retired Naval Captain and former Dean of Faculty and Academic Programs at the Industrial College of Armed Forces, touched me with the immediacy of his response to feedback on the volume. John dropped everything—in what is such a short summer here at NDU—to give me invaluable suggestions. I have the good fortune of working more directly with ICAF faculty than most people and John is such a tremendous example of what they offer.

Within my home institution, the National War College, I work with the top people in the field of professional military education. Each and every one of the women and men who walks in the door as a faculty seminar leader understands the value of strategic logic and the dangers of sloppy thinking. The complementarity of the faculty is an amazing phenomenon. It is a privilege to share the challenge with these professionals.

Every year, we educate roughly two hundred U.S. and foreign officers and civilians about the need, the development, the trade-offs, and the implementation of national security strategy. One of the unique aspects of the professional military education system of the United States, and especially of the National War College, is that each of us around our seminar tables learns, faculty and students.

I regularly learn so much from men and women, civilians and officers—foreign and domestic—who have been in the operational world. This experiential learning causes us to remember each and every day that national security strategy, in the end, is about decisions taken and implemented by women and men who often put their lives on the line because they are protecting our nation. This is true for the civilians who work in the area as much as, obviously, for those in uniform who are deployed around the world. I am about to welcome my fifteenth class at the college, meaning I have interacted with roughly three thousand officers. Some of my former students are in the highest levels of leadership to engage in strategy. Sadly, a few are no longer with us. In each and every case, I appreciate what I learned from these individuals about professional military education and there would be no point of this book without these students.

Several individuals at the college itself merit special thanks. The Commandant of the National War College, Major General Marné Peterson, U.S. Air Force, sets a wonderful tone each and every day. Her support and enthusiasm is crucial to our work. Captain Steve Camacho, USN, is a friend and valued colleague. His detailed feedback on the joint professional military education material, based on his service on the Joint Staff, was crucial to this volume; coming at a time when he was simultaneously Department of Military Strategy chairman and Director of Field Studies was a tribute to his professionalism and friendship. I hope I have answered his concerns adequately but any residual errors are mine alone. Drs. Roy Stafford and Dave Tretler, retired Air Force officers and superb educators, care deeply about teaching and professionalism, which is instrumental to our success as an institution. Col Mark Pizzo, USMC (retired), does all the hard stuff so we as faculty do not have to; we owe him a tremendous debt but all too often forget to tell him that. Dr. Joe Collins helped with resources. Col Mark Tillman, USA, and Col Kevin Keith, USAF, cheerfully answered a number of inquiries about their respective Services' PME. Ms. Susan Sherwood always asks incisive questions and offers some encouragement. Dr. Bud Cole is my best friend. As I have noted in other places, he has taught me so much over the years and never ceases to help me become a better person and scholar.

Introduction

Several months into the war in Iraq, a colleague in a traditional academic setting complained to me that she was disturbed that "the military guys running the war weren't educated enough to be doing this important business" for the world. I retorted that several of the senior officers prominently displayed on the nightly news were educated at my institution, the National War College, where they earned an accredited master's degree. The colleague was stunned. What I took from this conversation was not any surprise that she held this misunderstanding, but a conviction that the level of education for military officers had to be better described to society. There was virtually no way this woman, a product of the Vietnam generation, could have known how deeply embedded professional military education is in the United States military tradition.

Professional officers in the U.S. system are arguably some of the better educated members of any profession in the country. Having been a faculty member at the National War College of the National Defense University for well over a decade, I do not say that with false pride. The levels of education required for each Service are determined by that Service and the Goldwater-Nichols Military Reform Act of 1986 also put Congress into military education, not for the first time.

Much of the debate about PME today focuses on whether it is as cost-effective as possible and whether it is appropriate to the types of crises that U.S. military officers will face in the future. A considerable amount of the education occurs in a seminar setting, particularly for more senior or graduate level work because of its emphasis on active learning. In the 1990s, the push for "distance learning" that was growing in civilian education also took root in the debate regarding PME. Some administrators in the PME system, as true in civilian institutions, thought that distance learning might introduce the best example possible for a method of educating that would ultimately cost less. Occurring at the same time that administrators in the civilian world and the military realms were trying to find manners by which to save money in the face of escalating information technology costs, many hoped that DL would be a cheaper way to provide education of adequate quality. Some argued it would actually be better quality because it allowed the student to work at an appropriate pace. Sadly, DL has not been overwhelmingly successful in any venue. For the PME community, it simply did not catch on and did not provide nearly the cost savings that its advocates hoped.

Another big change in PME over the past two decades has been the growth of war gaming as a tool. War gaming is not new to the military's approach to thinking about potential conflicts around the world. One of the most famous features of the interwar years, from 1918 to 1941, was the Naval War College war gaming for conflict in the Pacific, known as Plan Orange, which began a startlingly accurate blueprint for the war that transpired between 1941 and 1945 against Japan in the Pacific theater.

PME has many common features with any other form of education, especially in the United States, but some important differences as well. Most faculty at PME institutions, especially higher-level programs such as "Senior Service" schools, are not tenured but are senior figures in their field, whether traditional academics, interagency specialists, or uniformed officers themselves. This means that they are accustomed to some of the uncertainty and constant change that characterizes the military's constant rhythm of change. Thus, the belief that the system must be frozen at any particular moment in time and not subject to change is far less common than in a system which does not value either current applications or quality education. In a system where rewards come from research and from the acclaim that refereed research entails for the institution, teaching quality is less important with a list of subsidiary effects. One of those effects is that research rather than quality of teaching becomes the measure of a faculty member's contribution to the institution. Teaching, in fact, may not seriously be evaluated because the tenure and promotion committee process may consider research to the virtual exclusion of teaching feedback from students. The quality of instruction as well as the coherence and timeliness of the course curriculum may suffer.

In the PME system, where the focus on the quality of the product—educating members of the national security community for perhaps immediate, direct application of acquired knowledge—is overwhelmingly weighed against any other activities, there is a need for constant, accurate assessment of the educational activity writ large. The military's tradition is to study its past actions to learn good and bad lessons, especially in preventing mistakes in the future that could be deadly for many; after each operational mission, the unit conducts a "hotwash" to evaluate the mission. The PME community firmly holds to serious, ongoing assessments of education technologies, techniques, curricula, and goals. As an example, the Joint Forces Staff College in Norfolk, Virginia, has a robust Assessment Program of more than a dozen full-time staff dedicated to nothing but assessing their curriculum, the teaching quality, variety of techniques employed, and all other aspects of their programs. In a dozen years of teaching at civilian institutions of higher education, my experience was that tenured faculty could completely ignore the results of any student feedback on teaching or on any class, should the faculty member seek to do so.

Another major difference between PME and traditional academic settings is the propensity in the former to accentuate current events. This does not mean that theoretical topics are not studied, but men and women from military backgrounds are conditioned to be aware of the situation confronting them as they decide on operational approaches to any possible scenario. This propensity reinforces the need for constant updating of courses but it does occasionally make students frustrated. No matter what the level of education, uniformed officers tend to want to become current to the exclusion of longer trends. Since students are in PME as a time-fixed "assignment" rather than an open-ended educational opportunity, the time issue becomes a zero-sum decision. By emphasizing current issues, something else is excluded from the educational experience.

Contrary to popular belief, one of the most cherished characteristics at PME institutions is academic freedom, protected through a policy of almost mythical importance called "not-for-attribution." Not-for-attribution guarantees students, faculty, and any outside visitors a commitment to make no direct association between the substance of any statement made in a seminar or lecture and the name of the speaker who made the statement. Absolutely fundamental to this policy being successful is the fidelity of all parties involved. In a system, however, where students are encouraged to think creatively even in the face of the hierarchical decisions that might be current policy to the contrary, students would not learn nearly as much if they had to worry that their promotions (or even careers) might be subject to termination for personal reasons or even insubordination (punishable under the Uniform Code of Military Justice) if they voiced their views in a manner for which they could be punished. In a not-for-attribution system, students and faculty of all types are protected by the peer pressure that the system imposes. Speakers are somewhat more willing to be candid in their lectures, meaning that students get a fuller understanding of the politics behind decision making as well as the goals of most programs. Finally, faculty can speak with the ability to play "devil's advocate," which may force students to think beyond their normal thought processes because faculty know they will not be quoted in newspapers as supporting or, more likely, criticizing existing policies that as untenured government employees they might be expected to agree with in a political environment.

Another interesting difference between PME institutions and civilian schools is the mix of faculty. At the Service academies, the mix is between tenured civilian academic professors and uniformed officers, often junior, who rotate through, often for three years, in their career path. At the more senior schools, civilian academics hired under Title X of the U.S. Code work with faculty on loan from various national security agencies of the U.S. Government and active duty military who are at various points in their careers. In the case of the "agency" and active duty faculty, these individuals are brought into the PME system because they offer years of expertise in areas where academics have often only seen the topic from a theoretical perspective. In this arrangement, few students in military education receive strictly military instruction, reinforcing the national belief system of civil-military ties.

Professional military education has been important to the United States and its armed services throughout its history. While General George Washington received his education by riding with the Redcoats in western Pennsylvania during the mid-eighteenth century rather than from a classroom in Carlisle Barracks, many of the General's troops had no education at all. These were the exceptions to the entire U.S. experience. One of the more successful of Washington's young general officers, Nathaniel Greene, had no formal military education as had several German officers who served—thus educated—those fighting for independence. Greene had read voraciously from the histories of other military figures, and then applied the lessons to the campaigns up and down the U.S. east coast until 1783 when the British were defeated.

During the administration of Washington's successor once removed, Thomas Jefferson, the fledging nation set up an academy for educating the army. The U.S. Military Academy at West Point, overlooking the Hudson River north of the growing New York City, has celebrated more than two centuries of educating young men and, after 1976, women to serve in the defense and national security of this nation. West Point, as it is colloquially known, has evolved in a variety of ways as has professional military education for the nation. Not only does West Point produce an officer corps in a much wider array of majors than the initial years when the nation needed engineers to help with nation-building across the vast continent, but it also produces officers with skills beyond those learned exclusively from books. A great emphasis that is ever more important is on representing the range of women and men, newly arrived and long here, who constitute the nation but have different physical attributes, religious commitments, ranging economic experiences, and speaking a number of languages along with English. This sort of transformation of the emphasis of the professional education is as profound as the changes that have rolled across this country in waves over the past two hundred thirty years.

The changes of the 1980s and Goldwater-Nichols, as the military reform of 1986 is known, had a profound impact on professional military education, too. While much of the changes were aimed at the final years of officers' active duty years, the impact rippled through the PME system because any emphasis on improving rigor had effects on all levels of education. In the 1980s, the shift toward a more rigorous academic content from a greater emphasis on developing collegiality was a reflection of the transformation of warfare for the United States along with an acknowledgement of the growing complexity of joint, interagency, and multinational strategic considerations in the aftermath of the Cold War. Further, the reforms to the military education system, taken in the context of broader changes to the military's role in providing national security to the nation, illustrated an understanding that the whole of the armed forces needed to change, according to the Congress. The push for reform did not come out of the blue but had evolved over a period of time, as was made dramatically clear by the problems of inter- and intraservice communications and coordination in the Grenada invasion of late October 1983.

The Services fought against the reforms, preferring to adhere to the existing and, as they argued, successful conditions in place. The Reagan administration also had serious doubts about the need for these changes but the Congress, in a bipartisan manner, worked to force reforms on a system that believed them unnecessary and actually damaging to security. The same reaction occurred to proposals to change professional military education, with the Services believing they knew what their officers needed, not outsiders who were not experts. Yet, twenty years after the Goldwater-Nichols law and subsequent educational reforms, known as the Skelton reforms, the U.S. armed forces are clearly more joint, more thoughtful, more integrated, and more interagency and international at a time when national security threats require such changes.

Professional military education is not the same as training or general education. Professional military education targets the fields which are crucial to the officer's specific rise to the upper ranks of service. While training is to convey a technical skill useful in accomplishing a particular task, education is intended to convey a thought process that can be applied in multiple circumstances. The PME system through which U.S. officers pass has the goal of educating its students to apply their analyses through a series of possible applications.

CHAPTER 1

The History of Professional Military Education

The history of military education within the United States is a rich and proud one. The Service academies, each with its specific aims and collective experience, have helped to create and solidify the Service cultures that remain crucial today. At the same time, the role of the academies has shifted within the Services as the Reserve Officer Training Corps assumed an increasingly crucial role, institutionalized as part of the professional military education (PME) program of the mid-twentieth century. Through the twentieth century and into the current decade, PME has increasingly been a method of bringing the Services together while creating some commonality in the ranks. Each of the Services has its particular concerns but increasingly PME is joint in nature, creating a more effective overall military force.

The Ground Force: The U.S. Military Academy

The U.S. Military Academy at West Point held its first courses in 1802, during the administration of President Thomas Jefferson, with its creation on the banks overlooking the Hudson River in West Point, New York. The Army's commissioning educational institution began with a heavy curricular emphasis on engineering, science, and mathematics, leading to a disproportionate percentage of its graduates in the nineteenth century receiving degrees in the hard sciences and engineering. This allowed the Army to produce several generations of officers who formed the basis of the Army Corps of Engineers. The Corps, in turn, was crucial to many of the infrastructure developments of the United States as it moved the frontier west and filled out national territory. With its blend of students from around the nation, nominated by their congressional representatives for admission, the Academy is one of the more highly competitive academic institutions for admission in the United States as are the other Service institutions.

Early Nonfederal Facilities and Military Education

In 1820, the American Literary, Scientific, and Military Academy began an innovative program with a military service commitment as well as liberal arts education in Norwich, Vermont. This body had a unique emphasis on the role of military education to prepare students for the possibility of serving the nation under arms. While this is the earliest instance in the United States of what ultimately became the Reserve Officer Training Corps (ROTC), it was a strikingly different model of accessing students into service while providing civilian education as well. The Academy in Norwich is now Norwich University, an educational location with a proud and extensive commitment to basic ideas of PME.

Two other institutions have developed similar precommissioning PME curricula but are not run by the federal government or the Services. These are the Citadel, located in Charleston, South Carolina, and Virginia Military Institute in Lexington, Virginia.

The Virginia Military Institute (VMI), the oldest state-supported military college in the nation, commenced in 1839 and has created a rich military tradition. One of the most famous graduates of the institution was General George Catlett Marshall, chief of staff of the Army during World War II. VMI offers a smaller number of major and minor fields for students to pursue on the grounds that this will generate higher quality rather than a watered-down curriculum. Slightly smaller than its sister institutions, VMI has a Corps of Cadets of just over 1,300 with students from seven foreign nations. Out of each year's graduating class, 40 percent pursue a commission in one of the Services.

The Citadel, also known as the Military College of South Carolina, began in 1842, built on the legacy of institutions charged with defending Charleston for twenty years. Although located in a city with a long naval tradition in the United States, the Citadel has provided a significant portion of Army officers who did not attend the U.S. Military Academy at West Point. The curriculum is similar to that at most undergraduate colleges with about eighteen possible majors and an appropriate number of minor courses to pursue. Like the Service academies in the United States, the Citadel emphasizes the whole development of students: physical, mental, leadership, and ethical-moral. Today, the Carolina Corps of Cadets numbers just under two thousand. The Corps includes women, as of the 1990s, and about twenty international officers pursuing U.S. entry-level PME. Roughly one of five graduates chooses to take a military commission upon completing the course at the Citadel.

Both the Citadel and VMI have highly selective admission policies and illustrate the draw that discipline and rigor along the lines of developing a career in the armed services offer youth today.

The Sea Services: The U.S. Naval Academy

The Naval Academy started its rich story in 1845 with its constitution in the Maryland capital city of Annapolis on the banks of the Chesapeake Bay. The Naval School, as originally known, started classroom education to guarantee not only fleet competence but also discipline. Prior to that, the U.S. Navy had taken the same path as the British Royal Navy of putting men to sea where they were bound to learn more applicable material than the French model of studying in the classroom.[1] Renamed the U.S. Naval Academy five years later, the school remains the entry-level education for Navy officers preparing to take a commission in the Navy or Marine Corps (primarily, but not limited to these Services) studying at the U.S. Government institution rather than a public or private university. Undergraduates at Annapolis are known as midshipmen. Graduates led the Navy as it took a greater role in beginning to spread U.S. power around the world. Nascent education at the Naval Academy showed the importance of tradition as well as skill in the fleet which has always been the hallmark of the U.S. Navy.

Army School of Application for Infantry and Cavalry

In the heady days of expanding the nation westward to the Pacific, the Army played a crucial role in that expansion. Roads needed building and communities, no matter how small, needed defense against attacks. Even wagon trains of families moving across the great open west feared moving without some sort of protection, which the cavalry was asked to protect. The location of Fort Leavenworth, Kansas, on the banks of the mighty Missouri River northwest of the growing Kansas City, became a crucial location for Army operations. Paralleling the initial path of the Oregon Trail to the northwest, Fort Leavenworth became the focus of several military activities and remains foundational to the Army of the United States today. In 1881, partially because of its location halfway across the country and partially because of fear that vast distances involved would fragment its force, the Army created the School of Application for Infantry and Cavalry at Fort Leavenworth. Two decades later, the school evolved into the General Service and Staff College. Eventually, it took its current title as the Command and General Staff College.

Similarly, Congress allocated $200,000 for the creation of the Cavalry and Light Artillery School at nearby Fort Riley, Kansas, in 1887. The school took several names over the first fifty years of the twentieth century: the Mounted Service School in 1907, then the Cavalry School immediately after World War I, in 1946 the Ground General School, and the Army General School in 1950. That educational facility ended in 1955 as PME in the Army became concentrated in Fort Leavenworth and Carlisle Barracks, Pennsylvania.

In many ways, establishment of these two schools aided in the creation of an esprit de corps. Yet, with the end of the nineteenth century, Army leadership, particularly under Major General Emory Upton, argued that a more systematic type of education would have to come about to raise the Army's ability to meet new foes as the nation evolved.

The Advent of Higher Level Education: The Naval War College

The Naval War College became an institution of education within the U.S. naval command structure in 1884 at Newport, Rhode Island. This grand setting overlooking the entrance to Narragansett Bay also began hosting the College of Naval Command and Staff in 1923 for its officers at the middle of their career path.

Founded by the Navy to bring up the quality of officers' thinking about the evolving role of the Navy in the nation's defense, two prominent individuals set in motion the thinking and ingenuity that has become the hallmark of the Naval War College. Commodore (ultimately Rear Admiral) Stephen B. Luce was the first president of the college and one of the four inaugurating faculty members was the son of West Point professor Denis Mahan. Also ultimately a rear admiral, Alfred Thayer Mahan was one of the original naval officers charged with developing a curriculum appropriate to the changing nature of naval warfare. The younger Mahan's lectures from the last two decades of the nineteenth century were consolidated in a book published under the title *The Influence of Sea Power Upon History, 1660–1783*. Still cited as a major work on naval issues, Mahan's work brought the Naval War College to prominence.

Elihu Root and the Foundations of the Army War College

Elihu Root, Theodore Roosevelt's secretary of war, sought to improve the Army leadership after the Spanish-American War of 1898. He proposed the creation of an Army General Staff, modeled on the German Army General Staff but this took several legal iterations to achieve. This move, with general officers at specified ranks on the staff, required an enhanced education by those officers. The Army Reorganization Act of 1901 raised the number of billets in the Army but did not deal with the technical changes confronting officers and enlisted personnel at the turn of the century. Later in 1901, Root convinced the Congress to authorize a War College Board which would create an Army education system and the formal study of military policy.[2] The General Staff Act of 1903 was the result of Root's efforts, with profound changes to the Army chain of command and leadership. The Army War College also began that year.

Additionally, Root was the driving force behind the conception of the Army War College. Begun in 1903 in Washington, D.C., by the Secretary of War, it became the highest level of Army education. The following year, President Theodore Roosevelt laid the cornerstone for Roosevelt Hall at the Washington Barracks, sited at the confluence of the Potomac and Anacostia rivers in southwest Washington, D.C., for the Army War College's building. After World War II, the college relocated north to Carlisle Barracks, Pennsylvania, as the National War College occupied former Army War College space in the nation's capital.

The Army War College concentrated its studies on the land power of the United States. As the threats and opportunities increased for the nation as its role in the world expanded, the Army arguably had the greatest challenge in understanding and applying the defense issues to its studies. The college also became an important site for the Service to reconsider its lessons to determine what could have been done better.

Finally, Root's impact as secretary of war also resulted in the creation of the Joint Army Navy Board, an amalgamation of senior officers from the two Services that functioned as an antecedent to the Joint Chiefs of Staff. For the better part of two decades, the Board coordinated the development of U.S. national strategy in conjunction with the secretary of war and the president. The Board was recreated in 1919 and became the Joint Board during World War II before it became the Joint Chiefs of Staff in the 1940s.

Between World Wars I and II: Marine Corps Education Comes of Age

For the United States, World War I represented a crucial departure from prior experiences. The U.S. forces did not enter the conflict until its final year but the PME system took a significant number of lessons from the experience. Mobilization of the force proved unsatisfactory and a school, the Army Industrial College, began in 1924. The Marine Corps noted many needs and began the establishment of several professional institutions. A markedly increased interest in an Army Air Corps resulted while the Navy became more careful in its planning and wargaming. The leadership of the U.S. military services came of age in the interwar years after 1918 and prior to the attack on Pearl Harbor in December 1941. While the United States had a significantly smaller role than the other states in the conflict in Europe between 1914 and 1918, many professionals in the U.S. armed forces found the U.S. mobilization and execution of the war less than optimal. As a result, the leadership who ultimately became crucial for the country in World War II, such as Ernest King and William Halsey in the Navy, Alexander Vandergrift in the Marine Corps, Hoyt Vandenberg and Hap Arnold in the Army Air Corps, and George Catlett Marshall, Dwight David Eisenhower, and Douglas A. McArthur from the Army, argued for the need to invest in education as the United States reexamined its experience in Europe. Significantly different education from the nineteenth century was the effect; arguably PME became the basis to what it is today.

Because it is the smallest of the Services, the Marine Corps is arguably the easiest at making large-scale changes to its education, among other things. In the years immediately following World War I, the Marine Corps began looking at education to make certain its forces were adequately prepared to determine the best courses of action, whether at the tactical or strategic levels. Major General

John Lejeune turned to education for all marines on the premise that was essential to the needs of the Corps.

As a result, the Marine Corps Officers' Training School began at Quantico, Virginia, in 1919. The emphasis in this education was on weapons and tactical use of those weapons. Brigadier General Smedley Butler pushed Lejeune's points further by creating two more courses, the Field Officers' Course in October 1920 and the Company Grade Officers' Course the following July. The three recently constituted courses became the basis to Lejeuene's "Marine Corps Schools" at Quantico.

Further into the interwar period, Marine leadership became convinced that studying amphibious warfare was the future of the Corps and incorporated that into studies. Under Brigadier General James Breckenridge, the Marine Corps curriculum changed dramatically to incorporate amphibious warfare along with close air support. The nature of Marine Corps Schools at Quantico was thus different from any prior period. The successes achieved during World War II a generation later were seen as based considerably upon these educational decisions during the interwar period.

Similarly, the Command and Staff colleges of the Services were all crucial to studying the changes under way in the military and global realms during this time frame. War college level education asked students to consider the options and challenges of conflict against a rearmed Germany or an increasingly militant Japan in the Pacific. The Navy War College assumed an almost mythical role in the history of U.S. PME because it had the foresight to engage in "war–gaming" against the Japanese in scenarios that became prescient for World War II in the Pacific theater. The officers in residence at Newport in 1934 discussed the challenges and operations of a four-year campaign through the islands of the southwest Pacific. One of the prouder compliments for PME in the United States came from Winston Churchill's assessment that World War II victory originated in the PME classrooms of the United States during the interwar period.

The roots of the Marine Corps Command and Staff College in Quantico, Virginia, were the Marine Corps Field Officers' Course started in 1920. Aimed at middle-grade officers, the Marine Corps Field Officers' Course showed these ranks how to conduct tactics and operations.

In the aftermath of World War I when the United States partially mobilized for conflict in Europe, the military leadership argued for the creation of an educational facility which would emphasize the logistics and mobilization functions within the Services. With the support of industrialist Bernard Baruch, the Army Industrial College commenced in 1924, the first college of its type in the world. One of its first student classes included Major "Hap" Arnold, later a founder of the Air Force. A graduate of the Army War College across the street, Major Dwight David Eisenhower, who later went on to use his education to orchestrate the European campaigns in World War II, served on the faculty in the 1930s.

An absolutely crucial aspect to PME was the National Defense Act of 1920. This law reorganized the Department of War, including creating an assistant secretary of war, which entailed having mobilization responsibilities to prevent the mess that preceded U.S. engagement in World War I. The first assistant secretary, appointed by President Warren G. Harding, was John Wainwright who tried to figure the best way to achieve the mobilization goal. One of the most prominent people who he consulted in his quest was financier and Wall Street icon Bernard Baruch. The consensus was that individuals needed education to approach the issue in a systematic manner but Wainright could not find funding for such a school before his replacement as assistant secretary in March 1923.

Wainright's replacement, Colonel Dwight Davis, U.S. Army (retired), believed that the school was even more necessary because he viewed the need for not only mobilization planning to prepare and execute a conflict but also for procurement that will provide the resources necessary to carry out the fight. His work and changing views in Washington made the funding accessible and the Army Industrial College received its Army mandate in February 1924. Throughout his life, Bernard Baruch, an instrumental supporter, reminded students that the Industrial College was partially to keep the armed forces "in touch with industry."[3] Two individuals early associated with the Industrial College were student Major H.H. "Hap" Arnold who studied in the 1920s and Major Dwight D. Eisenhower who studied and taught simultaneously.

Eisenhower worried about the disjunctures between the industrial sector of society and the military. During his 1931 lectures, he noted "a cleavage developed between those that plan for operations and that that would supply the efforts."[4]

One of the necessary alterations to the structure of the military command confronting President Franklin Roosevelt involved creation of the Joint Chiefs of Staff and theater commands.

Evolution in the Air Corps

With the role of air operations in World War I, the Air Corps began the decade after the conflict with its position embedded in the Army. The air forces remained part of the Army until the Defense Reorganization Act of 1947 when the separate Air Force began. The Air Corps Tactical School started in Montgomery, Alabama, at Maxwell Air Force Base. Eventually, this institution evolved into the Air Command and General Staff College and the Air War College, collocated with the current Air University.

The National War College: Jointness and National Security Strategy

Once the United States engaged in a conflict far away in an international environment, other professional military questions requiring education arose. Before World War II ended, the idea arose that operations across Services were going to be necessary. General Hap Arnold of the Air Corps oversaw the creation of the Joint Army-Navy Staff College which opened in 1943. The class sizes were quite small but others, such as Canadian and British observers, made obvious the need for a greater educational access. The end of World War II made clear that the United States armed forces, for all its brilliance, competence, and capability, still had a number of problems that Generals Dwight D. Eisenhower and George C. Marshall recognized could be obstacles for the future. In the summer of 1944, after the Normandy invasion, Lieutenant General Lesley J. McNair, head of the Louisiana Maneuvers and Education for the Army during the war, died after he parachuted into northern France. The assessment of the incident indicated that McNair died as a result of "friendly fire," resulting from poor coordination and communication across the Services. Friendly fire casualties were not new in World War II nor did they end after that war but Eisenhower and Marshall believed that an educational institution which promoted better synchronization across the armed services was needed.

Well before the war ended, the strategists in Washington were thinking about the postconflict educational system for the armed forces. The Joint Chiefs of Staff Special Committee for Reorganization of the National Defense became known as the Richardson Committee after its chairman, Admiral James Richardson, USN (retired). The committee held dozens of meetings but there were passionately held views on all of the options developed by the committee members. Ultimately, the leadership of the Services seemed to oppose each other on the most important aspects of the proposals set forth. The Congress held hearings on the various approaches, resulting in the Navy, Army, and Air Force divisions of the national security community laid out in the National Security Act of 1947. The next decision related to the ways to implement a more efficient PME system.

With this decision taken, the structure of the PME system was the next priority for decision. The Joint Army–Navy College commandant, Lieutenant General John Dewitt, USA, working with the recommendations of the Richardson Committee and others, such as Bernard Baruch, had the job of crafting a PME system. Among the recommendations were that the Industrial College become the Industrial College of the Armed Forces, not merely titled to the Army, and that the National Defense University (NDU) be created to "complete the formal education" for officers "at the 25-year level of commissioned service, of senior officers of the armed forces in military strategic war planning."[5] The range of educational questions asked was wide and the conclusion was not foregone.

President Truman chartered the "Gerow Board" under Army Lieutenant General Leonard T. Gerow to consider the issues of joint education. The outcome of the study was to recommend five separate colleges, each with a joint orientation at a National Security University in the nation's capital. In addition to the Industrial College and the National War College, the proposal included an Administrative College, a Joint Intelligence College, and a State Department College. The site recommended was at Fort Lesley J. McNair where the Army War College had been located since Roosevelt Hall opened in 1903. The Gerow recommendations included the suggestion that the Army War College classes remain suspended as they had during World War II.

Even with the specter of the Cold War on the horizon, the majority of the Gerow recommendations were rejected due to lack of funding. The Army Industrial College became the Industrial College of the Armed Forces and the National War College was born.

The National War College opened its doors to the first class of officers in September 1946. The college, with its early and enduring mission of educating the future leadership of the United States in national security strategy (instead of topics of one Service), has always had a "joint" tradition. This allowed all Services and several civilian agencies to have students and faculty among those at the college and to participate in the study of national security strategy. In 1947, the college's first international affairs advisor, Ambassador George Frost Kennan, formerly in the Soviet Union, penned the "Mr. X" article in *Foreign Affairs* magazine, providing the basis to the "containment" doctrine that the United States and its allies used against the Soviet Union from 1947 through 1989 when the Berlin Wall fell. The international affairs advisor has been the second ranking officer for much of the National War College's history, representing the State Department's commitment to the need for jointness and better understanding of military-civilian coordination. Other initial instructors included Bernard Brodie and Sherman Kent. The National War College, sixty years after its creation, still concentrates exclusively on joint PME at the national security strategy level.

Armed Forces Staff College

The Armed Forces Staff College, aimed exclusively at intermediate-level officers, began in 1948 in Norfolk, Virginia. As true with the National War College, U.S. military leadership was concerned that mid-grade officers were not aware enough of the support and advantages open to them by utilizing all forces instead of those exclusively of the officer's Service. The Armed Forces Staff College role from the beginning was to ameliorate that condition. Five years after the NDU commenced in 1976, the Staff College entered its structure as the intermediate-level education facility for jointness. In 2002, the college took the new title of Joint Forces Staff College to emphasize the joint role in its education and philosophy.

The Advent of Air Force Education: The Air Force Academy

The Army retained control over the air component until the Defense Reorganization Act of 1947 that created the separate Air Force and a number of other changes in the national security structure of the United States. In establishing a new Service, the Congress also authorized creation of the Air Force Academy, putting it in Colorado Springs, Colorado. This institution, along with sister schools West Point and Annapolis, had strong emphases on math, engineering, and high codes of moral ethics in preparing the corps of the various Services. As the twentieth century progressed, the Academies admitted women in 1976 and opened educational possibilities for studies in fields beyond engineering and military science.

During the late 1950s, President and former general Dwight D. Eisenhower attempted to reform the military in ways that some believe today was the precursor to Goldwater-Nichols forty years later.[6] Eisenhower's experience as chief of staff of the Army after World War II indicated his understanding that waste, redundancy, and poorly coordinated leadership plagued the U.S. Army and, by extension, the entire U.S. military. Eisenhower sought to reform the military to prevent the problems he saw but was thwarted by the Services in conjunction with the Congress at a time when the fear of the Soviet global expansionism outweighed any desire to minimize, if not eliminate, the conditions which undercut efforts at jointness and education activities.

Reserve Officer Training Corps and Twentieth-Century Changes

The traditional accession route for military officers was through the three Service academies for much of the history of this country. In the middle of the nineteenth century, the United States began to require a military education component tied to the land-grant colleges being established throughout the country by Congress. The 1863 law, the Morrill Act, became the basis for the development of the ROTC system which began in the first decades of the twentieth century. Between 1863 and the end of the first decade of the twentieth century, land-grant schools offered a form of military education that was deemed sufficient for those who eventually entered the ranks of the military officers of the nation but those numbers were quite small. The overwhelming majority of officers received their commissions because of their educations at the Academics.

The period before the United States joins World War I in 1917 is replete with many competing pressures in the military realm. One of these is a significant press by individuals such as Emory Upton and Grenville Clark to promote private military education in camps outside of the Service academies and the land-grant colleges promoted by the Movvill Act. This movement to promote citizen-soldiers became known as the Plattsburgh Movement because some of the training was done at facilities outside of Plattsburgh, New York. This attempt to move initial military professional education outside of its traditional sources was an important evolution into new parts of society. The ROTC programs did not begin formally until the National Defense Authorization Act of 1916 that established Army ROTC programs while the Navy's were set in 1926 and the Marine Corps' program six years later.

Turbulence in the Vietnam Era

PME faced the same upheavals that faced society in the 1960s and 70s. The U.S. military came to the end of the Vietnam experience in a somewhat tattered position with illegal drugs, racial turmoil, and challenges to its ethical fabric coming from the ranks. President Richard Nixon's 1973 decision to create a volunteer force instead of relying on a conscription system generated further considerations. Many societal pressures were encouraging the admission of women to the Service academies and into increasingly integrated roles in the Services, which bothered many traditionalists. Attendance at Senior Service schools gave the future leaders of all the Services the freedom to think through some of these questions while also addressing the top level of national and Service-dependent security questions.

One of the major areas of disagreement related to the role of the mandatory ROTC programs hosted at many universities around the nation. The education had evolved past the land-grant aspect of the Morrill Act a hundred years earlier, but educational institutions often required ROTC as part of a civic duty in exchange for education. Many youth and faculty members who disagreed with the premises of U.S. involvement in Southeast Asia protested the mandatory participation for all men on campus and often objected to the military's simple presence at many schools. As a result of the controversies, ROTC was discontinued in some schools and made into voluntary participation as the military draft ended in 1973 replaced by the All Volunteer Force. While there are still some concerns about military programs on campus, the self-selection process that is the basis of the system has largely ended the controversy generated by mandatory systems.

One of the other contentions about ROTC programs revolved around the rigor of the academic content. As the Cold War continued, military officers increasingly sought to build a bond and loyalty needed on the battlefield while civilian faculty challenged programs to have greater precision in thought and content. Many ROTC officials simply could not understand why civilians did not grasp the need for the bonds that save buddies on the battlefield instead of "book learning" that may have no application in the world. The solution became a mix between the sides that characterizes ROTC today but did raise the important question of how stringently military officers were thinking through the problems facing them.

Along with promoting thought about the reforms necessary to improve the Army and other Services in the aftermath of Vietnam, the idea of military officers earning advanced degrees also took root as a given in the military system. In the year the war ended (1973), a quarter of the officer corps in the United States earned graduate degrees. By the mid-1990s, that figure was thirty-eight percent with the lieutenant colonel/commander and colonel/captain ranks having an extraordinarily high percentage of those graduate degrees.

One of the last great reforms resulting immediately after the Vietnam War was the decision, implemented in 1976, to admit women to the Service academies. The controversy and fury that this evoked among the student and alumni bodies of the Academies were swift and intense but not unlike that suffered by single-sex institutions such as the University of Notre Dame when it took the same decision in 1973. Many traditionalists simply did not want to see changes to a system believed to work well as a single-sex institution. In the case of the Service academies, Congress could impose the change to student body composition with the passage of legislation, even if there was plenty of op-ed page bluster. But, the controversy did not end with simply changing the number of men versus women admitted to a class. The implication for the Services, as they greatly altered the character of the ranks, was important for warfighting, camaraderie, and all other aspects of the newly emerging All Volunteer Force. But, it was this precisely new force that meant the admission of women could not be delayed further. The controversy about women in the armed forces (and thus PME) has not ended but ebbs and flows, depending on other issues under public debate. Even though women were admitted to the Academies in 1976 and their numbers rose as a result of choosing the All-Volunteer military as a career, women were seriously constrained for many years from participating in specialties that would put them into combat situations. This question remains a hotly debated one and probably will remain such. In terms of education, the combat prohibitions limited the paths that women could take immediately upon graduation and, in reverse order, their chosen major fields.

The Air University

With the creation of the Air Force in 1947, the need for an educational establishment to push the new Service with innovative theory and practice was obvious. Actually prior to the 1947 Act, a small cadre of thinkers to consider air and space aspects of warfare had been convened. They built upon the prior Army Air Corps schools at Maxwell Air Force Base in Montgomery, Alabama. The Air University with many component parts reflects the Air Force interest in education, pioneering tactical and operational thinking, and the bonding of a new Service. As the time has proceeded, the United States Air Force has gradually filled in the pieces to an extensive and graduated PME culture.

The National Defense University

In the mid-1970s, Congress authorized creation of an umbrella organization at Fort Lesley J. McNair in Washington. Home of the two "joint" Senior Service colleges, the National War College and the Industrial College of the Armed Forces since 1946 when they were constituted as they currently exist, Fort McNair was a military reservation with one of the few open spaces within the District of Columbia for expansion. The NDU often sees itself as the chairman of the Joint Chiefs of Staff's educational institution because of the prominence of JPME at NDU. It is often called the "Chairman's University" to denote the ties between the joint Services, the chairman of the JCS, and the University. The intent behind the creation of the NDU was to remove the administrative burden from the two colleges in hopes of allowing them freedom to focus on education. It has, however, grown dramatically beyond its initial scope. Along with the colleges, the NDU included a strategic concepts center that became known as the Institute for National Strategic Studies in 1984.

Expansion of NDU began with a greater pace in the 1990s and 2000s. In the middle of the 1990s, an educational institution with emphasis on the evolving role of information in national security began at NDU: the Information Resources Management College. The Information Resources College is not technically part of the PME system because it is not mandated under the OPMEP, but it is seen as useful by many senior officers worried that the United States is vulnerable to cyber attacks.

NDU took on a number of new missions during the 1990s, largely due to the national policy of "engagement" with foreign states, in this case with foreign militaries. The Center for Hemispheric Defense Studies for the western hemisphere, the Africa Center for Strategic Studies, and the Near East/South Asia Center for Strategic Studies all began, each with a mission tailored to the region each represented but generally aimed at bringing democratic values and practices to states that have suffered from poor governance or were transitioning to democratic systems.

As the nation became more aware of the need for homeland security and for issues relating to nation-building and reconstruction operations, the NDU became an obvious location for much debate about how PME should proceed in the United States. Some within the Defense Department argued that the education should continue as it has for generations because the nature, character, and conduct of war are enduring phenomena; others pushed for a radically new emphasis on terrorism and counterterrorism, while still others thought that the crux of the education should look to future efforts for nation-building while understanding the lessons of stability operations from prior experiences. Indeed, the 2006 Quadrennial Defense Review (QDR) discussed making the NDU into the National Security University, although doing so would require much more than simply renaming the establishment. The 2006 QDR also encourages the development of new aspects of education within the Services to promote cultural and language awareness, which may have contributed to difficulties in Iraq and Afghanistan in the early years of this decade.

1980s: "Final Straw," Revising the Joint Chiefs, and the Push Toward JPME and Upgrading of PME

The Vietnam War had shown the drawbacks of poor coordination across the Services and the dangers of inter- and intra-Service rivalries as well as financial expenses generated. The response to the *Mayaguez* seizure in May of 1975, where the coordination was also lacking at a time of national self-doubt just days after the fall of Saigon to the North Vietnamese forces, reinforced concerns. This was a long process that had begun as early as the end of World War II when Eisenhower and Marshall acknowledged the need for the Services to better understand themselves and their peers in the other Services to avoid mistakes. Some thirty-five years later this was still true as the Iran Hostage rescue disaster (April 24, 1980), resulting in the death of eight personnel in the desert, had shown how vital communications and coordination were.

The major push to shift to a mandatory joint PME experience resulted to an extent from the late October 1983 invasion of Grenada, seen as the final straw in a two-decade fight to get the Services to operate more jointly.[7] The disastrous October 1982 bombing of the Marine barracks in Beirut, during a controversial intervention in the Lebanese civil war, reinforced the view that the necessary inter-Service communication within the EUCom chain of command organization was broken. Grenada was a small Caribbean island with a leftist government that President Ronald Reagan believed threatened the security of U.S. medical students on the island as well as the stability and freedom of the island, and it became a battleground as 1983 unfolded. The controversial New Jewel Movement, led by Maurice Bishop, was overthrown on October 19, resulting in considerable instability on the island. In the last week of October, U.S. forces, with support from the prime minister of nearby Dominica, moved into Grenada under "Operation Urgent Fury" to rescue the medical students and install a more predictable, pro-Western regime to govern Grenada. President Reagan also clearly was more comfortable putting a pro-Western government in charge of Saint George's than one teetering on the brink of a leftist ideology.

In an engagement where U.S. military superiority was absolute, the interoperability of U.S. forces proved shockingly poor. Communications between the Services was poor, if not non-existent in some cases, leading to deadly situations for U.S. forces as well as sheer redundancy of effort in an era when the United States could afford redundancy. Mercifully, fewer than twenty U.S. personnel died but the bureaucratic competition seemed intractable as "Operation Urgent Fury" was reviewed. Coming two days after a completely different event, the bombing of the Marine Corps barracks in Beirut, Lebanon, on October 23, 1983, scrutiny was heightened for all U.S. military operations. For Congress, however, the redundancy displayed during Urgent Fury was not seen as a benefit but an expensive complication. The Grenada experience added tremendous emphasis to the push already under way in Congress to promote greater "jointness" across the U.S. armed services.

Both the uniformed and civilian leadership reacted strongly to this shocking series of events. The chairman of the Joint Chiefs of Staff, Air Force General David Jones, had begun arguing for a radical change in the structure of the Joint Chiefs in 1981, much to the consternation of many fellow senior officers. By the discombobulating in Grenada, lines were drawn between those who opposed reform and those who believed that Jones had spoken the truth. While there were many on each side, one of the major critics of this state of affairs was the senior senator from Arizona and former presidential candidate, Barry Goldwater, a retired Air Force reserve general. Goldwater felt that the military leadership was too concerned with protecting its turf instead of protecting the sons and daughters fighting for the nation. He used Grenada, as did many of his congressional colleagues, as an opening to force the leadership of the Services to come to grips with the unmanageable system in place.

The other named cosponsor of military was a Democratic representative from Alabama, Bill Nichols. Nichols joined Senator Goldwater in demanding, in a series of hearings in the mid-1980s, that the Army, Air Force, Navy, and Marine Corps get into closer coordination to achieve a synergy of force, instead of providing repetitive, redundant defense capabilities for the nation.

In his volume, *Victory on the Potomac: The Goldwater-Nichols Act Unifies the Pentagon*, James Locher describes the long struggle for reform against tremendous pressure from cross-cutting groups in the Services: some civilians within the Defense Department, many of the various chiefs of staff of the Services over the period from 1981 to 1986, some retired officers of note, and a variety of congressional members and staffers. But, Locher illustrates that the power of the trend in problems facing the nation, in combination with congressional concerns about funding and capabilities, led to the success of reformers.

One of the goals other critics had in mind for military reform was to bring the traditionally independently minded Navy into line with the other Services. The Navy leadership, with its proud traditions of individual captains running their ships rather than having the operational hierarchy that exists in the Marines or Army, for example, did not value classroom, joint military education to the same degree as did the leadership of the other Services. In the Navy, "school house" discussions could never match the education provided by being out on deployment with the fleet. As both the Army and Air Force increasingly made opportunities available for their officers to earn master's degrees, for example, the Navy approach was to send its officers to the Naval Post-Graduate School in California, for advanced technical degrees, rather than simply advanced degrees in any field as was true in the other Services. The Navy also had less interest in pushing its students to attend joint PME schools. Goldwater-Nichols, the shorthand often used for the 1986 law that mandated greater commitment to jointness, sought to force the Navy to accept education for its forces while promoting the benefits of encouraging more stringent academic standards. In February 1985, the Center for Strategic and International Studies, a think tank with strong respect in the military community, unveiled its study, *Towards a More Effective Defense*, recommending a high component of PME reform.

The Goldwater-Nichols Military Reform Act of 1986 was the first major revision of the U.S. defense community in the era of the All Volunteer Force. The goals of passing this set of reforms were much broader than merely aiming to alter PME. The intent of creating legitimate, enduring joint interoperability was paramount but many other things also were intended.[8]

Marine Corps University

At the same time, there was a push within the Congress to create a more rigorous PME system for the Senior Service level PME curricula. As noted, the Marines had begun taking steps to create a more professionalized educational program as early as 1919. Seventy years later, the Marine Corps University emerged as a consolidated, integrated program for marines at all stages in their professions. The consolidation of schools at the Quantico Marine Base south of the nation's capital made for a somewhat more coherent package of schools, although the obvious synergies of education have never been proven.

The 1990s and Civilian Accreditation

As early as the 1970s, moves to affect the curriculum of two PME institutions—the Industrial College and the National War College—had been under way. The Clements Commission on Excellence in Education, headed by Deputy Defense Secretary William Clements, sought to unify the two curricula while bringing them to an even higher level. One outcome of the Commission was to create the NDU that opened in 1976. Located at Fort Lesley J. McNair in southwest Washington, DC, where ICAF and NWC had been for decades, the purpose of NDU was to assist the overall education at these two primary schools. Thirty years later, the Industrial College has expanded responsibilities, to include the creation of a number of smaller programs at McNair but the Industrial College and National War College retain their distinct missions. In fact, the Industrial College's responsibilities have grown considerably by its inclusion of the Defense Acquisition University to focus on acquisition function within logistics and mobilization. In the mid-1990s, all Senior Service schools began the process of achieving nationally recognized academic accreditation. Prior to that point, many graduates of Senior Service colleges referred to their time at War colleges as "sports scholarships" or "gentlemen's years," implying that the academic standards were not particularly high. Congress increasingly in the 1980s began to challenge the assumption that the nation's national security community could afford such a luxury of a year off.

In the mid-1980s, two complementary trends began that were intended to raise the quality of PME at the Senior Service school levels. Both of these trends were primarily promoted from outside of the colleges. First, the colleges began hiring civilian academics as long-term faculty under Title X of the U.S. Code. There had long been traditional, nonmilitary academics on the faculties of the colleges but they were a decided minority of the faculty. In fact, the bulk of the educators who taught at the War colleges were active duty, uniformed officers, and agency representatives whose participation mirrored the relative number of participants from their home agencies. This meant that all War college students had some State Department or intelligence community faculty members and students for each class. By hiring more Title X faculty, the expectation was that these more traditional educators would inculcate their culture, teaching methods, and academic standards into the curricula of the schools.

The second trend was for the Senior Service colleges to acquire accreditation. Pushed largely by Senator Sam Nunn, Democrat of Georgia, and Missouri Democratic representative, Ike Skelton, the idea was that if the PME curricula of the

Senior Service colleges met the standards set forth by civilian, professional accrediting bodies, the rigor long desired for these programs would be guaranteed. The accreditation process which takes a number of years to accomplish was achieved by the six Senior Service schools by the end of the 1990s.

Concerns About Service Academy Education and Ethics

The manner by which one would attain increased rigor in the curricula included changing the faculty composition at the accession stage of PME: the Service academies. While the Naval Academy had a balance between civilians and uniformed officers that mirrored each other, the Air Force and Army induction institutions did not have the same support for civilian participation. The Military Academy and Air Force Academy both supported a faculty composition considerably weighted toward military personnel because the esprit de corps of the Services was seen as more important than the spirit of intellectual curiosity that pervades any civilian academic campus. In 1993, however, Congress determined that the imbalance, with 97 percent of the faculty in military uniform at the Military and Air Force academies, could not continue.

At the same time, a series of controversies at the Service academies raised concern about not only the intellectual rigor of the institutions but the ethics training and education of the men and women attending these schools. While these issues continue well into the first decade of the twenty-first century, each mirrored similar problems in civilian society and illustrated that the U.S. armed forces is populated by men and women like those from each and every community in the nation.

In terms of PME, however, each of the Academies has had an enduring controversy which it sought to address through altering the educational messages sent to its students. Beginning in the mid-1990s, the Naval Academy faced a series of ethical challenges that were not only violations of the Honor Code but obviously legal and chain-of-command violations. The problems involved sexual harassment, a car-theft ring, and even questions of murder. On top of the Navy's residual damage done because of the Tailhook scandal in 1991,[9] the Navy's hierarchy sought to stem the damage being done to the Service by the growing perception through the 1990s that the Navy held ethical standards below those of society in general.

The response was swift and layered. Navy curricula for all classes of midshipmen began to more thoroughly integrate ethics into discussions. Additionally, the Navy established a center for ethics at the Academy in 1998. While some outsiders challenged the worth of such commitment to the overall education system, misdeeds by midshipmen slowed dramatically as fewer cases appeared.

The concerns at the Air Force Academy were equally alarming to some in the public and again resulted in changes in educational emphasis. In the early 2000s, a series of women who had attended or were still attending Colorado Springs came forward to charge they had been raped by their peers but had been either silenced or ignored by the Air Force Academy leadership, proving, by extension, that the Air Force condoned rape and did not offer equality for women at the Academy. After initial denials, the entire leadership of the Air Force Academy was relieved of command, replaced by men and women officers who vowed to change radically the ethical and sexual treatment standards at the school. This did not change the effects for those who had gone before (in many cases forcing young women to abandon their hard-fought admission to this elite institution) but the situation has calmed considerably.

Immediately after the Air Force seemed to address the atmosphere which tolerated harassment, questions of religious tolerance arose in Colorado Springs. In the middle of the first decade of this century, the question arose of whether the Air Force Academy (and the Service as a whole) endorsed fundamentalist Christianity over the tolerance of other sects and religions. The controversy is broader than merely the education at the Academy but the exposure of the young cadets to religions different from their own or from a single view evokes strong concerns across all sectors of U.S. society and threatens to detract from the education of officers for service to the nation.

The Military Academy has also been involved in the controversy about sexual harassment and other ethical questions over the past decade. None of the Service academies have been immune to the debates and traumas that have arisen across society. Additionally, the cadets and midshipmen at all of the Academies have had to ask questions about how the wars in Iraq and Afghanistan have been conducted.

The Era of Joint and Coalition Warfare: The Need for Pinnacle

With the Clinton and Bush administrations, U.S. military practices increasingly included not only joint (more than one U.S. Service involvement) but also combined, multinational, and coalition forces to carry out an operation. Afghanistan has proven a perfect example as Navy SEALS worked jointly with Army Rangers in the initial operations (2001–2003) but the mission has increasingly been turned over to NATO forces in a coalition activity. Yet many senior U.S. flag officers have never studied such operations, much less engaged in them. Beginning in 2003, the NDU as the school for the national strategic missions of the nation started offering a new level of education called Pinnacle. The shortest of any PME courses, merely five days, Pinnacle is oriented toward the new challenges that are organizationally and operationally the most complex possible. Administered by the NDU in its Norfolk facility near the Joint Forces Command in that area, Pinnacle is largely an exercise to illustrate the complexities and challenges of the new environment. The course's short duration manifests both the pace of the activities in which two- and three-star officers operate as well as the time that any commander has to make decisions. After only three years, it is impossible to measure the course's success in altering major aspects of PME.

The Future of PME: Professional Security Education?

For many, the idea that PME can evolve easily into professional security education is an obvious, easily accomplished change. For others, the idea of sacrificing the hard issues of traditional defense studies for the sake of incorporating new security concerns is sacrilege. PME, with each word having an important relevant emphasis for individuals engaging in it, requires that every additional topic will force something else out of the curricula because PME is a finite experience for each program. Determining what is to be excluded for the addition of something else will be an extraordinarily important and controversial exchange which the United States has not yet completed.

Notes

1. John Yaeger, *Congress Influence on National Defense University* (Washington, DC: George Washington University, 2005), p. 28.

2. Terrence J. Gough, "The Root Reforms and Command," accessed at http://www.army.mil/CMH-pg/documents/1901/Root-CMD.htm.

3. *The Retired Officer*, May 2005, accessed at http://www.moaa.org/magazine/May2005/eisenhower.asp.

4. Ibid.

5. Ibid.

6. Greg H. Parlier, "The Goldwater-Nichols Military Reform Act of 1986: Resurgence of Defense Reform and the Legacy of Eisenhower," written at Marine Corps Command and Staff College, Quantico, VA, 1989, accessed at http://www. globalsecurity.org/military/library/report/1989/PGH.htm.

7. For detailed studies, see James Locher, *Victory on the Potomac* (College Station, TX: Texas A&M Press, 2004) and Gordon Nathaniel Lederman, *Reorganizing the Joint Chiefs of Staff: The Goldwater-Nichols Act of 1986* (Westport, CT: Greenwood Publishing, 1999).

8. There have been several reviews of Goldwater-Nichols to celebrate its fifteenth anniversary and more will appear to commemorate the Act's twentieth anniversary in 2006. Locher's *Victory on the Potomac* is one but others include Lederman, *Reorganizing the Joint Chiefs of Staff*.

9. Tailhook is a nongovernmental association of naval aviators who held their annual conferences in Las Vegas. Seen as a networking group, questions arose about sexual harassment of women officers and other "conduct unbecoming of an officer" at these meetings for which members had received government time and sanction to attend. Ultimately the Navy, after much public embarrassment, restructured the time associated with the Tailhook annual conferences to reduce the overindulgence in alcohol and raw behavior.

CHAPTER 2

Curricula and Institutions

Curricula for Professional Military Education

Professional military education (PME) in the United States has several different stages. Chart One, from the National Defense University (NDU), indicates a graphic for how this spreads over an officer's career. The levels are Precommissioning (undergraduate) PME, Intermediate-level PME, Senior-level PME, and Flag/General Officer PME. Embedded in all levels is an ethical and moral dimension as well as increasing emphasis on "jointness" to fulfill the requirements of the Goldwater-Nichols Military Reform Act of 1986 which set certain legal requirements for officers to progress through the ranks. Yet, it is important to note that an officer will not achieve a high enough rank to engage in joint professional military education (JPME) as described in the *Officer Professional Military Education Policy* ("the OPMEP," as it is known) if she/he does not excel at the lower rank learning that one must master in the individual Service. This chapter will consider these levels through the institutions applying various curricula to the issue.

JPME and the OPMEP

The Officer Professional Military Education Policy, or OPMEP, gives specific guidance on the PME, such as the level at which these individuals learn and the type of courses they take. The OPMEP is directive in terms of guidance on faculty deemed acceptable to teach JPME, the ratio of faculty from one Service, the student-to-faculty ratio, learning methodology, and other mechanical issues that are determined by the marketplace in civilian academic settings. These guidelines guarantee a consistency across Services in developing JPME.

To clarify these issues while setting the tone for the PME institutions, the OPMEP guidance appears here in an excerpted form.[1]

1. Precommissioning. Military education received at institutions and through programs producing commissioned officers upon graduation [West Point, Annapolis, Colorado Springs, and ROTC programs around the nation along with Federal and Staff Officer Candidate Schools (OCS) and Officer Training Schools (OTS)] ... [The] focus [is on] preparing officer candidates to become commissioned officers within the Military Department that administers the precommissioning program. The curricula are oriented toward providing candidates with a basic grounding in the U.S. defense establishment and their chosen Military Service, as well as a foundation in [joint warfare].
2. Primary. Education typically received at grades 0-1 [Second Lieutenant in the Army, Air Force, and Marines while Ensign in the Navy] through 0-3 [Captains in the Army, Air Force, and Marines while the rank is called a Lieutenant in the Navy].
3. Intermediate. Education typically received at grade 0-4 [Major in the Marines, Air Force and Army and Lieutenant Commander in the Navy].
   a. JPME Phase I (Services). Service ILCs [Intermediate Level Colleges] teach joint operations from the standpoint of Service forces in a joint force supported by Service component commands.
   b. JPME Phase II. The Joint and Combined Warfighting School at the JFSC (Joint Forces Staff College) examines joint operations from the standpoint of the CJCS [Chairman of the Joint Chiefs of Staff], the JCS [Joint Chiefs of Staff], a combatant commander and a JTF [Joint Task Force] commander. It further develops joint attitude and perspectives, exposes officers to and increase their understanding of Service cultures while concentrating on joint staff operations.
   c. JAWS. Provides a separate, single-phase JPME curricula reflecting the distinct educational focus and joint character of its mission ... The Services may recognize JAWS as an intermediate-level or senior-level PME equivalent.
4. Senior education for ranks 0-5 [Lieutenant Colonel in the Air Force, Army, and Marines and Commanders in the Navy] and 0-6 [Colonels and Captains for their respective Services]
   a. JPME Phase I and II (Service Colleges). Service SLCs [Senior Level Colleges] provide JPME Phase I and in-resident JPME Phase II education. Service SLCs address theater- and national-level strategies and processes. Curricula focus on how the unified commanders, Joint Staff and DoD use the instruments of national power to develop and carry out national military strategy, develop joint operational expertise and perspectives and hone joint warfighting skills.
   b. JPME Phase II. JCWS (Joint and Combined Warfighting School) at JFSC (Joint Forces Staff College) provides JPME Phase II for graduates of JPME Phase I programs to further develop joint attitudes and perspectives, joint operational expertise and hone joint warfighting skills.
   c. JAWS provides a separate single-phase JPME curricula reflecting the distinct educational focus and joint character of its mission. The Services may recognize JAWS as an intermediate-level or senior-level PME equivalent. JAWS meets policies applicable to intermediate- and senior-level programs.

    d. NWC [National War College] provides a separate, single-phase JPME curricula reflecting the distinct educational focus and joint character of its mission.

    e. ICAF provides separate, single-phase JPME curricula reflecting the distinct educational focus and joint character of its mission.

    f. Advanced Joint Professional Military Education (AJPME) builds on the foundation established by the institutions teaching JPME Phase I.

5. G/FO [General/Flag Officers] JPME prepares senior officers of the U.S. Armed Forces for high-level joint, interagency and multinational responsibilities.

In the beginning of the JPME era, Phase I and Phase II were treated quite differently. Both the junior and senior courses at all institutions other than the National War College and Industrial College provided only Phase I education, because the institutions were so heavily associated with their individual Services. Graduates of these schools were required to attend the Armed (now Joint) Forces Staff College in Norfolk, Virginia, to qualify for the necessary Phase II certification. Revisions in the curricula of these institutions over the years, however, have allowed the Senior Service schools (Army War College, Navy War College, Marine War College, and Air War College) to convey appropriate Phase I and II certification upon graduation. As noted, the Industrial College and National War College, by the entirely joint nature of their curriculum, have always granted Phase I and II certification on their graduates.[2]

The JPME certifications that officers seek in the United States are classified as Phase (or Level) I and II. These refer to the levels of education required for students to become joint specialty officers (JSO), which was mandated by the Goldwater-Nichols reform. The OPMEP, most recently issued on December 22, 2005, stipulates that JPME has "continuum and flow" and involves five levels of education.

1. Preparatory JPME taught during precommissioning [undergraduate] and primary [Service specific competency such as the Marine Basic School] schools.

2. JPME Phase I [is] taught at Service intermediate-level colleges (ILC) and Service senior-level colleges (SLC) in-residence (for programs that have not been accredited for JPMEII) or as a Distance Education (DE) or Distance Learning (DL) option.

3. JPME Phase II [is] taught at the Joint Forces Staff College and Service SLCs.

4. The separate single-phase JPME programs at the National War College (NWC), Industrial College of the Armed Forces (ICAF) and Joint Advanced Warfighting School (JAWS).

5. G/FO [General/Field Officer] courses.

As specified in the PME requirements set forth by the OPMEP, these standards and preparation are required for advancement and promotion. This will be explored in the chapter.

The Context for Professional Military Education

Noteworthy for those who have never served in uniform is understanding that all students at PME institutions have a required height/weight/physical fitness standard and so their days also contain a specific period dedicated to physical training, or "PT." While not specified in the PME requirements, these standards and preparations period go unsaid but are absolutely required.

One fundamental difference between PME and civilian institutions is the selection process. The overwhelming student base at PME schools is active duty, uniformed officers. In each institution, those students do not apply for admission but are selected by their Services and by allocation determined by the Joint Chiefs of Staff or the individual Service leadership. The active duty students are not self-selected as they are at civilian universities. PME is part of the career process as well as important for the promotion process; self-selection would not work in this system but the decision on the part of the Services sends a message to the students about the Service's view of a student's promotion feasibility. The civilian members of classes, however, may apply for PME or may be selected by their home agencies but they are chosen by the institutions. The uniformed students, thus, have little voice in the school where they receive PME education.

One new phenomenon that assists in the education process is the proliferation of foundations at the PME schools. These private foundations, usually 501 (c)(3) under the IRS Code, allow the colleges and universities flexibility in funding some activities for which tax dollars cannot be used. This funding allows students the opportunity to socialize with some extraordinary speakers or to engage in some activities for which federal spending cannot be used while private money can be spent.

Beginning in the 1990s, PME institutions at all levels have sought and achieved, in most cases, civilian education as well as professional military accreditation. The undergraduate schools, the Naval Academy, Air Force Academy, and Military Academy, have long met the requirements for outside accreditation because their degree-granting status would disappear without serious peer respect. For other PME institutions, the civilian accreditation began with congressional interest in the subject in the late 1980s. The accreditation is conveyed by professional associations such as the Middle States Commission on Higher Education, Accreditation Board for Engineering and Technology, the Committee on Professional Training of the American Chemical Society, and the Higher Learning Commission of the North Central Association of Colleges and Schools. The military accreditation is conducted by the Services that oversee the development of curriculum at each institution and the J-7 division of the Joint Staff which provides independent outside evaluators to satisfy its requirements.

While PME institutions are military schools, thus subject to military rules of order, chain of command, and other governing principles, the point of professionalizing the curricula is to raise the standard of thinking, education, and overall learning experience. At the same time, adopting these more traditional civilian measures and evaluations, the institutions increasingly adopt not-for-attribution and nonplagiarism norms to their policies. These are intended to guarantee that the PME institutions will foster open thinking for the purpose of developing necessary thinking skills.

The institutions listed here are listed by their commonly used designators. The Naval War College, for example, is roughly equivalent to the NDU in terms of what it controls within the Navy's sphere of education compared with what NDU controls. Because the component parts are within the Naval War College, however, everything is listed as such. In the Army structure, however, the Command and General Staff College and Army War College are not collocated nor in a chain-of-command relationship and so they are listed separately. The listing is in alphabetical order.

The Air Force Academy

USAFA, Colorado 80840
Voice: 719.333.1110
Electronic mail: scbw_all@usafa.af.mil
Web site: www.usafa.af.mil

The introductory level educational institution for the Air Force is the Air Force Academy in Colorado Springs, Colorado. In the rare air of the eastern slopes of the Front Range south of Denver, the cadets enter the educational system of the Air Force for a four-year undergraduate experience. The Academy began in mid-1955, some eight years after the Air Force itself resulted from the Defense Reform Act of 1947. The Air Force Academy teaches cadets the procedures, values, culture, and goals of the Air Force while also providing them with a strong undergraduate education. While many students were originally engineer majors, the Academy (along with its naval and army counterpart institutions) now provides a balanced curriculum that encourages a variety of educational paths. The entrance requirements are stringent and exacting, with many entering plebes both outstanding athletes and scholars from their high school backgrounds. The Air Force Academy also has research facilities, such as the Institute for National Strategic Studies, which conducts studies of topics particularly germane to airpower but not limited to that topic.

The Academy is a four-year school accredited by several civilian professional organizations. It is also central to the entry of Air Force officers to the Service.

The Air Command and Staff College

225 Chennault Circle
Maxwell Air Force Base, Alabama 36112
Voice: 334.953.6494
Web site: wwwacsc.maxwell.af.mil

The Air Command and Staff College qualifies for what is known as intermediate-level PME, aiming at officers in their middle levels of their careers (roughly 13–14 year mark), with the bulk of the students either at the rank of major or selected for major (majors select). Located at Maxwell Air Force Base in Montgomery, Alabama, the college replaced the Air Corps Tactical School which operated there between 1931 and 1942. The college, assuming its current title and form in 1962, dedicates the studies of its graduates to the requirements and opportunities for air and space operations in a combined and joint atmosphere. The students spend ten months studying a curriculum that spans many areas of concern to the Air Force. Working from a seminar-based residential curriculum but with a parallel distance learning program, the Air Command and Staff prepares students for command-level jobs while reinforcing thinking on the security challenges facing the nation and acclimating students to the needs of joint and combined operations in future activities. The residential program fulfills Joint Professional Military Education (JPME) Level I requirements under Goldwater-Nichols. Graduates receive a Master of Military Operational Art and Science degree, as of 2000. The distance learning portion of the curriculum has existed since 1949 and offers six courses and a number of exercises for the students who cannot spend time in residence in Montgomery, but students in this option do not earn a master's degree.

Air and Space Basic Course

Maxwell Air Force Base, Alabama 36112
Web page: www.asbc.maxwell.af.mil

The Air and Space Basic Course is the entry-level PME program of six weeks' duration for second lieutenants to learn the most fundamental material on how the Air Force engages in aerial conflict. Relatively recently, the course has added a five-day cooperative educational experience with the Senior Noncommissioned Officer Academy which is collocated at Maxwell. The basic course offers a focus on practical, hands-on educating in leadership, communication, and the interaction between officers and the enlisted personnel they will command.

The Air University

Chennault Circle
Maxwell Air Base, Alabama 36112
Web site: www.au.af.mil

Orville and Wilbur Wright taught flying at a school in Montgomery which became Maxwell Air Force Base and became the center of airpower studies in the 1930s with the location of the Army Air Corps Tactical School. Dating to 1946, the Air University is headquartered at Maxwell Air Base in Montgomery, Alabama, although it also has some coursework at Wright-Patterson Air Force Base in Dayton, Ohio, and some across town in the Gunter Annex in Montgomery. As part of the Air Education and Training Command, Air University is a multi-level institution that runs from the senior most educational activity within the Air Force for its officers on air operations and air power strategy to precommissioning coursework. The components of the Air University include the Air War College; the Air Command and Staff College; the Squadron Officer College; the College of Aerospace Doctrine, Research and Education; the Ira C. Eaker College for Professional Development; the School for Advance Air and Space Studies; the Air Force Officer Accession and Training Schools; and the Community College of the Air Force. Several of these are not strictly speaking PME but part of the Air Force educational structure. Because the Air University manages the College for Enlisted PME, the Air Force Senior Noncommissioned Officer Academy is also part of this responsibility as is the Air Force Institute for Advanced Distributed Learning.

Air War College

325 Chennault Circle
Maxwell Air Force Base, Alabama 36112-6427
Voice: 334.953.6093
Facsimile: 334.953.7225
Web page: http://www.au.af.mil/au/awc

For the Air Force, the Air War College is the Senior Service school in PME. Its students are senior officers who look at the use of air and space resources at the strategic level with a great emphasis on joint and coalition operations. The Air War College also has an open enrollment program for its nonresident senior program, the only one in the PME system. The nonresident program does not award accredited JPME credit. In any year, the Air War College educates 245 resident students from the United States and more than 6,000 nonresident students from a range of federal agencies, all of the uniformed services, and foreign nations. The Air War College received its certification as a Phase I institution in 1989 and will stand for Phase II certification in fiscal year 2007.

Armed Forces Staff College

See the Joint Forces Staff College.

The Army Command and General Staff College

Fort Leavenworth, Kansas 66027-1352
Voice: 913.684.5604
Web site: www-cgcs.army.mil

The Army Command and General Staff College, located at historic Fort Leavenworth in northeastern Kansas, is the intermediate-level PME institution for the

Army. Founded in 1881 as the School of Application for Infantry and Cavalry to help enlisted personnel keep their eyes aimed at their missions, the school evolved into the General Service and Staff College in 1901, and then its current title later in the twentieth century. Aimed at field grade (majors and major-selects) officers, the "Command and Staff," as it is frequently known, offers students an education on Army—joint, combined, and multinational—operations across the range of threats. Students are predominantly U.S. Army, but there are also a significant number of foreign officers on exchange from their nations as well as officers from other Services of the U.S. force. The bulk of the students are resident for their coursework, although there are some distance learning offerings. Additionally, at Command and Staff, two other functions coexist: the School for Command Preparation and the School for Advanced Military Studies. The School for Command Preparation offers weeklong courses targeting four different levels of command preparation. These bring leadership and management skills to all prospective Army commanders. The School for Advanced Military Studies, on the other hand, "educates and trains officers at the graduate level in military art and science to develop commanders and General Staff officers with the abilities to solve complex military problems in peace and war" (CGSC Web site). Along with the educational institutions, other resources are available at Fort Leavenworth to assist in the development of Army leaders. These include the Center for Army Lessons Learned, the Combat Studies Institute, and the Combined Arms Library. The Army also publishes *Military Review* from the Fort Leavenworth facilities.

Army War College

U.S. Army War College
Attn: Public Affairs Office
122 Forbes Avenue
Carlisle, Pennsylvania 17013-5234
Voice: 717.245.3131
Electronic mail (public affairs): AWCC-CPA@carlisle.army.mil
Web page: http://www.carlisle.army.mil/usawc/daa/external_site/collegehome.shtml

The Army War College, an outgrowth of Secretary of War Elihu Root's major reforms to the Service at the beginning of the 1900s, began in Washington, DC, but has been in Carlisle Barracks, Pennsylvania, for almost sixty years. The college has six components under its leadership, beginning with the Army Senior Service College which began in 1903. In the residential program, lieutenant colonels and colonels of the Army, and their peers in smaller numbers from the other Services, spend ten months at the college studying the operational issues and national strategy for ground forces. The nonresidential program offered is of two years' duration with periodic two-week visits to Carlisle. A major portion of the time spent at Carlisle Barracks for this course is dedicated to the study of strategic leadership. The Army War College received its certification for Phase I JPME in 1989 and expects to achieve Phase II accreditation during fiscal year 2007. The college has also achieved accreditation by the Middle States Association for Higher Education. The other organizations are the Strategic Studies Institute, which is a research institution concentrating on ground force issues among a long list of topics, and two centers with specialized interests, namely, the Strategic Leadership Institute and the Peacekeeping and Stability Operations Institute, which provide the Army with opportunities for study, dialogue, and research on the topics. The other two portions of the college are the Army Physical Fitness Institute to study physical conditioning for Army personnel (with broader applications across the Services) and the Military History/Army Heritage and Culture Center, with its emphasis on expanding the understanding of the college's role in Army history. The Army War College not only educates students and civilians but is also a focus for conferences within the Army community. Much of its work is in the outreach field to expand contact between the public and the Service personnel. The Army War College quarterly publishes the professional journal, *Parameters*, on research relating to relevant topics.

Capstone

Building 59, Room 280
408 4th Avenue, S.W.
Fort Lesley J. McNair
Washington, DC 20319-5062
Voice: 202.685.4260/2330/2332
Facsimile: 202.685.4256
Web site: www.ndu.edu/capstone

The Joint Chiefs of Staff, under General John Vessey, USA, charged the NDU with creating a PME curriculum for newly chosen flag and general officers which became pilot programs offered in 1983 and 1984. CAPSTONE was initially a voluntary, eight-week course, but was reduced to six weeks in 1986. In calendar year 1987, CAPSTONE changed to a required course offered four times annually to meet the requirements of the new Goldwater-Nichols Military Reform Act of 1986 for all newly selected general and flag officers. CAPSTONE has a focus different from other PME programs in several ways. The rank of participants is higher while class sizes are significantly smaller than those of other programs. Additionally, the course examines U.S. force deployment in combined and joint operations to maximize national security policy requirements. The program also allows newly selected flag and general officers to meet combatant commanders and to interact with four-star retired officers who participate in the program as Senior Fellows, either in Washington or during its travel periods. Particularly in light of the emphasis on creating a more "joint" military after Goldwater-Nichols, this type of education stresses joint and unified operations between and across the Services. CAPSTONE focuses the new flag and general officers on the issues that relate to these changes in the Services' and U.S. military force's concept, such as the greater use of joint task forces, multinational operations, and the role of interagency operations in homeland security missions.

The mission is to "review[ing] the elements of national power . . . integrated to achieve national security objectives." The flag and general officers examine six learning objectives: unified, joint, and multinational operations; the national security environment, intelligence support system, joint and individual Service doctrines and capabilities and planning processes; force acquisition concerns and the implications of conducting unified, joint, and multinational operations under the current conditions; integrating military strategy with national security strategy; the civil-military relationship in the United States as well as interactions with Capitol Hill; and understanding force protection/risk management concerns. CAPTSONE accomplishes its mission through lectures, national and international travel, discussions with local civilian and military leaders, and participating in a crisis decision exercise.

Center for Strategic Leadership Studies

Air War College
Maxwell Air Force Base-Gunter Annex
Web page: http://leadership.au.af.mil/sls-abt.htm#history

Established in 2002, the Center for Strategic Leadership Studies at the Air War College desires to become the "preeminent institution for the study of senior military and government leadership" (http://leadership.au.af.mil/sls-abt.htm#history). The Center conducts research through individual concerns and collaborative efforts in the U.S. PME system as well as in conjuction with foreign scholars. Particularly noteworthy about this Center is an extensive Web page dedicated to leadership resources, which is accessible at http://leadership.au.af.mil/sls-ndex.htm.

College for Enlisted Professional Military Education

Maxwell Air Force Base-Gunter Annex
Web site: www.cepme.maxwell.af.mil

The College for Enlisted Professional Military Education, located at Maxwell Air Force Base Gunter Annex, is part of the Air University. Its mission is to instill in enlisted personnel a sense of how these individuals are part of today's Air Force. The courses falling under this college include the Airman Leadership Schools, Noncommissioned Officer Academies, and the Air Force Senior Noncommissioned Officer Academy.

College of Naval Command and Staff

See Naval War College

College of Naval Warfare

See Naval War College

General Service and Staff College

See Army Command and General Staff College.

The Industrial College of the Armed Forces

Eisenhower Building
Fort Lesley J. McNair
Washington, DC 20319
Web site: www.icaf.ndu.edu

The Industrial College of the Armed Forces is the oldest JPME Senior Service school, dating back to 1924. Its roots were as the Army Industrial College, founded from the World War I experience that the United States was poorly prepared to mobilize and "resource" operations that contributed to winning World War I in Europe. The industrialist Bernard Baruch was instrumental in the creation and constant review and revision of the curriculum. One of the first faculty members at the college was a young Dwight David Eisenhower in the early 1930s. As World War II unfolded and the needs of a truly joint force became more important, the Industrial College took its new name. ICAF developed the specific niche of studying the adequate resources and mobilization process to accomplish its national security goals. In 1976, with the establishment of the NDU at Fort Lesley J. McNair in southwest Washington, DC, the Industrial College became one of the two anchor schools at that institution. In the 1990s, the Industrial College also took on the mission of studying the process of acquisition, with a specific concentration within the college on acquisition. Today, the more than three hundred annual graduates of the college, coming from the Services, various U.S. government agencies, international students, and a handful of civilian industrial students, earn a master's degree in National Resource Management. The Industrial College looks at resources for the strategic level but also has a significant commitment to the acquisition process for the U.S. government's military arm. ICAF students travel, as part of their required curriculum, to study industrial bases in the United States and abroad. The ten-month curriculum offers a Level II JPME certification for the military students as required under Goldwater-Nichols legislation in the 1980s.

The Joint Forces Staff College

7800 Hampton Boulevard
Norfolk, Virginia 23511-1702
Voice: 757.443.6086

Facsimile: 757.646.6026
Web site: www.jfsc.ndu.edu

During World War II, it became clear that the overwhelming majority of officers in the United States had no idea how to think or interact in what we now refer to as a "joint" manner. With military campaigns in two distinct theaters thousands of miles apart, the leadership of the U.S. armed forces pushed for education that would encourage officers to think in the manner most needed by the campaigns. This led to the creation of the Army and Navy Staff College (ANSCOL) in 1943 at Washington Barracks in the southwestern quadrant of the District, on the third oldest active duty post in the nation. With the end of the War, the ANSCOL also ended. A new institution and curriculum under the chief of naval operations' oversight on behalf of the Services began on June 28,1946, initially in Washington, DC: the Armed Forces Staff College. Six weeks later, the college moved to Norfolk, Virginia, because it was accessible to the new class of 150 officers from all the Services, where it remains today at a site that had been the U.S. Naval Receiving Station. The initial commandant, Lieutenant General Delos C. Emmons, U.S. Air Force, opened the first class on February 3,1947 with faculty who had served in all parts of the World War II Force. The Armed Forces Staff College joined the NDU, although it was always a PME institution, in 1981. Originally aimed at intermediate-level officers, the college became certified to grant Phase II PME accreditation in 1990 under the Process for Accreditation of Joint Education for JPME. It is also accredited by the Middle States Association for Higher Education.

The Joint Forces Staff College has a wide array of courses and programs, aiming at the intermediate-level officers of the various Services. With a stunning range of acronyms, the emphasis for the programs is on aspects of warfighting. The programs offered at JFSC are

- Joint Advanced Warfighting School (JAWS)
  The emphasis in the Joint Advanced Warfighting School is on coordinating and planning the most thorough integration of warfighting skills from across the Services. Its students, majors/lieutenant commanders and lieutenant colonels/commanders, are instructed on the "art and science of joint, interagency, and multinational planning and warfighting at the strategic-operational level of war." (http://www.jfsc.ndu.edu/schools_programs/jaws/overview.asp) This goal, set forth by the CJCS *Officer Professional Military Education Policy*, fulfills required coursework for JPME levels I and II while the college is also pursuing accreditation for a master of science degree in Joint Campaign Planning and Strategy. The methodology of teaching is for senior officers from across the Services and civilian academic professors to teach twelve students in one of two seminars with equal representation from across the Navy, Army, Air Force, and Marine Corps. A main message for graduates is the importance of Operational Art in joint affairs. JAWS is an eleven-month program.
- Joint and Combined Warfighting School (JCWS) JPME-II. The stated goal of the JCWS is to develop "*[A] Joint and Combined Operational Warfighter who is able to significantly contribute to the development of comprehensive plans and effective execution across the range of military operations (ROMO)*" (http://www.jfsc.ndu.edu/schools_programs/jcws/overview.asp). The mission is "*To educate military officers and other national security*

leaders in joint, multinational, and interagency operational-level planning and warfighting, to instill a primary commitment to joint, multinational, and interagency teamwork, attitudes, and perspectives" (http://www.jfsc.ndu.edu/schools_programs/jcws/overview.asp).

- Joint Command, Control, Information Operations School (JC2IOS). The emphasis for this school is Joint Command, Control, Communications, Computers, and Intelligence Operations. This course emphasizes operations at the operational level.
- Joint Planning Orientation Course (JPOC). A brief course for officers and civilians at the operational level on deliberate and crisis planning.
- Joint Transition Course (JTC). This is a course to level the field on knowledge about issues such as joint planning, Service perspectives, DoD planning, and other issues a student would have learned in a JPME Phase I course. This is aimed at students from abroad, civilians, and uniformed personnel who had not had the opportunity to achieve Phase I certification.
- Reserve Component Joint Professional Military Education (RC JPME). This is a course for the Reserves and encompasses many things that this portion of the military have not had the opportunity to consider.
- Electives
- Senior Fellows
- International Officers Program
- Homeland Security Planners Course (HLSPC). This is a course to emphasize the lessons learned about homeland security over the past few years as well as the emerging threats.
- Joint, Interagency, and Multinational Planner's Course (JIMPC). This is a certificate course for senior uniform and civilian officials on several levels of planning.
- Office of Force Transformation. This is an office to make certain the lessons of "transformation" of the U.S. military are inculcated into the students at all levels.

The Marine Corps University

Quantico Marine Corps Base, Virginia
Web site: www.tecom.usmc.mil

The Marine Corps University is the youngest of the PME universities with its history only beginning in 1989. The history of Marine Corps education, however, is much richer and dates back a century further to 1891 with the founding of the Marine Corps School of Applications. The best known commandants of the Marine Corps, such as Major General John Lejeune, pushed for the study of tactical employment of weapons and other aspects of military education that were bound to improve the quality of Marine Corps' officer leadership. The first of many programs was the Marine Corps Officers' Training School at Quantico. Further institutions were the "Marine Corps Schools" with the Field Officers' Course, the Company Grade Officers' Course, which are the bases for the Marine Corps University. The Marine Corps takes pride in offering courses to make certain the new environments confronting the Corps were always addressed. Amphibious warfare, for example, became a study topic in the 1930s under Brigadier General John Breckenridge. As the Cold War ended and the Services appreciated the growing complexity of the issues confronting U.S. forces in a world of no other superpower, the Marine Corps created the three-level professional education structure.

In 1964 the Senior Course was reconstituted as the Command and Staff College, while the Junior Course became the Amphibious Warfare School. Seven years later the Staff Noncommissioned Officer Academy commenced at Quantico. The Noncommissioned Officer Basic Course started in 1981, followed by the Senior Course for Staff Sergeants. The University itself resulted in 1989 with five degree-granting programs and several programs. The Marine Corps University received accreditation in 1999 to grant a master's degree of military science to graduates of the Command and Staff College, while the War College and School of Advanced Warfighting degrees received accreditation in 2001 and 2003, respectively. The Amphibious Warfare School merged with the Command and Control systems courses to form the Expeditionary Warfare School in 2002. The General Officer education system resulted from changes to the Senior Leader Development Program a year later.

The Marine Corps War College

Quantico Marine Corps Base, Virginia
Web site: http://www.mcu.usmc.mil/MCWAR

Awarding a master's degree in Strategic Studies, the Marine War College is located at Quantico Marine Corps Base, an hour south of Washington, DC. Named the College of War Studies with its founding in 1990, this is a small, select school of roughly eighteen students from the Marine Corps and other Services who study in conjunction with civilians relevant to the national security community. The Marine War College concentrates on war and operations other than war in a joint, combined, and multinational environment. This ten month program fulfills the Level I requirement for JPME under the Goldwater-Nichols Military reforms of 1986; at the end of 2006, it will also award Level II certification. The curriculum at Marine War College has four components: War Policy and Strategy, National Security and Joint Warfare, Regional Studies, and General Studies.

The Military Academy

West Point, New York 10996
Web site: www.usma.edu

Created by President Thomas Jefferson on March 16, 1802, the U.S. Military Academy at West Point is the oldest military educational institution in the nation. Many of the most famous and recognizable names in U.S. military history graduated from the Military Academy: Robert E. Lee, George Pickett, Douglas A. McArthur, Dwight D. Eisenhower, and John Abizaid. Many of those graduates actually returned to teach the following generations. Charged with preparing the future leaders of the Army through precommissioning education, West Point, as it is commonly called, offers its students a firm educational basis balanced between the sciences and the arts. Originally, West Point produced primarily engineers who often built the nation as it expanded westward. The Corps of Engineers, and the Military Academy itself, was prominent in the settlement of the nation as well as the Army as a Service. Eventually, however, Army students got the option to elect social science and humanities majors on the premise that the Army needs the widest range of expertise in this complex world. The education at the Academy seeks to enhance four aspects of the cadet's experience: the academic experience, the physical challenges to keep fit for service, the moral-ethical questions facing and to face the officers of the future, and the military culture to indoctrinate incoming officers about the Force they will join upon graduation. The faculty of the institution are a mix of civilian and serving officers. Many of the civilians are retired officers but this is not a requirement for job application. Instead, the Academy values high quality teaching and an awareness of the research ongoing in the faculty member's field. The Academy graduates roughly a quarter of the Army's new lieutenants, or approximately 900 cadets, annually. Those who graduate have experienced an intense academic program while also experiencing an active life on campus where leadership and character traits are developed. The Commission on Higher Education of the Middle States Association of Colleges and Schools accredits the Academy as meeting educational standards of institutions in the region while several departments are individually accredited as well.

The National Defense University

George C. Marshall Hall
Fort Lesley J. McNair
Washington, DC 20319
Web site: www.ndu.edu

Created in 1976 to reduce administrative responsibilities for the Industrial College of the Armed Forces and the National War College, in the 1990s the NDU spread its interests into broader issues. By the first decade of this century, the NDU has poised itself to begin anew as a much expanded institution that may reach far beyond PME into other fields. The NDU itself is not an educational institution but an umbrella organization with responsibilities to the chairman of the Joint Chiefs of Staff. Headed by a three-star active duty officer whose Service affiliation rotates among the Marine Corps, Air Force, Army, and Navy, the University originally had only three notable components: the Industrial College, the War College, and the research arm known as the Institute for National Strategic Studies; the Armed Forces Staff College was a part of the University located at the Norfolk, Virginia, campus three hours south of the nation's capital. Gradually other programs such as the North American Treaty Organization (NATO) Staff Course for all officers who will serve at NATO headquarters in Belgium began to move to NDU as a logical, central location for such an activity. Early attempts at making sense of the changing role of information in professional military concerns led to the creation of the Information Resource Management College in the mid-1990s. Similarly, goals of trying to expand U.S. engagement and support for civilian over military rule led to the founding of the Center for Hemispheric Defense Studies (1997), the African Center for Defense Studies (1998), and the Near Asia/South Asia Center for Strategic Studies (1998); all are located at Fort McNair, although the Africa Center did much of its initial work on the African continent. In 2005, all of the Centers left NDU control but remain components of the Department of Defense and they are still located at the NDU campus at Fort McNair. Toward the end of the 1990s, the Clinton administration supported arguments that not only did military officers need PME but their civilian counterparts would have benefited from understanding the internal and international environments for national security strategy. It created the Defense Leadership and Management Program that is now the School for National Security Executive Education, although the curriculum has been through considerable upheaval over the time of the course. In the early part of the current decade, NDU also created the Center for Technology and National Security Policy for research on technology interacting with national security and military policies; the Center for the Study of Weapons of Mass Destruction that grew out of a prior emphasis on counterproliferation; the Institute for National Strategic Studies that considers items requested by the Joint Chiefs of Staff as well as self-generated and contains the National Strategic Gaming Center; and the Institute for Homeland Security. The NDU Press began publishing *Joint Force Quarterly* during General Colin L. Powell's tenure as chairman of the Joint Chiefs of Staff in 1992. As the increased emphasis in the government is on the evolving issue of homeland security, there have been indications that some congressional members would like to reformulate NDU into the National Security University but that has not occurred. Similarly, the 2006 *Quadrennial Defense Review* discusses the NDU becoming the National Security University but this is not yet the case.

The National War College

Roosevelt Hall
Fort Lesley J. McNair
Washington, DC 20319-5078
Voice: 202.685.4343
Facsimile: 202.685.4654
Web site: www.nwc.ndu.edu

The National War College, located in the historic Roosevelt Hall where the cornerstone was laid by President Theodore Roosevelt in 1903 for the Army War College, has been the Senior PME school for national security strategy since its creation in 1946. Founded because of Generals Eisenhower and Marshall having concern about the knowledge that the Services had of each other and their civilian counterparts, the National War College has always concentrated the educational experience on preparing future leaders of the national security community on the rigorous logic necessary for developing and executing national security strategy for the nation to accomplish its strategic interests. The college has fought efforts to dilute this mission through adding other courses to the detriment of this rigorous thinking. Two hundred students, 75 percent from the Services (including international officers issued invitations by the chairman of the Joint Chiefs of Staff) and a quarter civilian, study at the college for a ten-month period that earns them a master's degree in National Security Strategy. Students are at the commander or captain levels (or lieutenant colonel/colonel equivalents). The college is unique in two major ways from all other PME schools: it has been a "joint" institution—whether across the Services or the civilian-military student body—since the beginning and the Department of State has played a role in the organization's leadership since Ambassador George Frost Kennan was the first international affairs advisor in 1946 as he wrote the "Mr. X" article that appeared in *Foreign Affairs* and became the basis to the "containment" strategy against the Soviet Union. The National War College has given itself the moniker of "Chairman's School" in PME.

The U.S. Naval Academy

121 Blake Road
Annapolis, Maryland 21402-5000
Web site: www.usna.edu
The secretary of the Navy, George Bancroft, created the U.S. Naval Academy in historic Annapolis, Maryland, forty-five miles east of the nation's capital, in 1845. A four-year program ranging from "plebe summer" for the incoming men and women who have been selected in a highly competitive process to participate in the Brigade of Midshipmen to the graduation. Upon commissioning, the Academy contributes a significant number of the officers to the nation's Navy and Marine Corps. The education, accredited by civilian accrediting bodies, is both technical and ethical, rigorous but diversified. The students must not merely do well in classroom activities but must engage in athletic competitions which prove their satisfactory preparation for the possible grinds of military service. The Academy has survived many challenges to its role in the PME. The changes brought by Navy aviation and submarine warfare, as well as new types of surface vessels, altered the composition of naval forces, and thus affected the curriculum. Similarly, the onset of the United States as a colonial nation at the end of the Spanish-American War of 1898 brought the Navy to prominence, especially in the Pacific basin. The Brigade of Midshipmen changed with the first African-American cadet, John Henry Conyers, arriving in Annapolis in 1872. It was not an easy transition but a continual one. Annapolis became gender integrated in 1976 when the first class of women arrived in Annapolis. The U.S. Naval Academy is considered the precommissioning stage of PME as are the other Service academies. The faculty at the Academy is a mixture of civilians hired under Title X of the U.S. Code and active duty navy officers who teach while also acting as role models for the student body. The Academy has professional accreditation as an institution from the Middle States Association and several of its individual academic departments are independently evaluated by their peers.

The Naval Post-Graduate School

Public Affairs Office—code 004
1 University Circle
Monterey, California 93943
Voice: 831.656.2023
Facsimile: 831.656.3238
Web site: www.npgs.edu

The Naval Post-Graduate School, known as the "PG School," was established at Annapolis in 1909 as a marine engineering school from which the Post-Graduate School evolved. The move from Maryland to the Pacific coast occurred in 1951 when the School was relocated "lock, stock, and barrel" to its new home. There have been many suggestions to close the Post-Graduate School over the years but none have succeeded because the Navy persuaded its critics that the education achieved at its institution was specifically tailored for naval officers and the security of the nation. The School has four programs under its purview. The Graduate School of Business and Public Policy concentrates on business pursuits and the impacts on public policy debates. As the Navy has altered its balance of officers, it has seen the need for officers with a greater business acumen grow. The School of Graduate International Studies considers issues in the complex global community, meshing them with traditional security studies' questions. The Graduate School of Engineering and Applied Sciences represents the heart of the PG School tradition, with marine engineering, naval engineering, and other applications for an officer afloat. The fourth school is the Graduate College of Operational and Information Sciences where computer sciences, operational research and other wide-ranging research topics appear as the regular curriculum. Students at the PG School may acquire their Level I JPME certification through distance learning conducted by the Naval War College, Army War College, Marine Corps War College and the Air Force War College. The Post-Graduate School also offers a range of courses for naval officers such as an Executive MBA and there are a series of research institutes which complement the PME coursework. These institutes include sponsored research programs, integrated graduate research and education programs, and institutionally funded research programs. Specific institutes include the Cebrowski Institute for Information Innovation and Superiority, in commemoration of the late innovator, Vice Admiral Arthur Cebrowski; the Modeling, Virtual Environments, and Simulations Institute; and the Wayne Meyer Institute of Systems Engineering. The School of Graduate International Studies has three research programs: Civil-Military Relations, National Security Affairs, and Defense Resource Management. The Graduate School of Engineering has research options to include Applied Math; Electrical and Computer Engineering; Mechanical and Astronomical Engineering; Meteorology; Systems Engineering; Oceanography; Physics; and Space Systems. In the Graduate School of Operational and Information Sciences, the four research programs include Computer Science; Defense Analysis; Informational Sciences; and Operations Research.

The Naval War College

Newport, Rhode Island 02841
Web site: www.nwc.navy.mil/defaultf.htm

The Naval War College is an institution of higher education with its emphasis on the Navy and aspects of naval warfare. Housed on Coaster's Harbor Island at Newport, Rhode Island, where it was established on the site of a former Rhode Island asylum, the Naval War College started in 1884 as an advanced institution for the study of navy warfare. Its founding president was Commodore Stephen Luce, and Rear Admiral Alfred Thayer Mahan, one of the prominent—if not the most prominent—U.S. scholars ever on naval issues, was one of four initial faculty members. Luce pushed for Naval War College curriculum to encourage a fundamental grasp of the nexus between technology, the Navy and national strategy. When Mahan replaced Luce as the president, he raised the visibility of the institution even further, largely because of the prominence of his own scholarship. His lectures were consolidated into a single volume, *The Influence of Seapower upon History, 1660–1783*, published in the early 1890s with long-term effects on U.S. strategy. At the beginning of the twentieth century, the college began promoting war gaming as a method of better studying possible naval warfare. Closed during World War I, the Naval War College remained open during the World War II when several of its graduates proved crucial to the war efforts, especially in the Pacific. By the late 1940s, the predecessor volume to the *Naval War College Review* was raising international attention to the thinking ongoing at Newport. In 1956, the Naval Command College was created to educate seasoned international naval officers. Similarly, in the 1960s two other programs began which have profoundly affected the Newport campus. The College of Naval Command and Staff for mid-grade officers emphasized tactics and operations. The College of Naval Warfare concentrates senior officers on strategic and administrative problems that will concern them in their future jobs. In 1972, Admiral Stansfield Turner, another innovator who was considered extreme by many, not only dramatically altered the curriculum but created a middle grade course for international officers in the Naval Staff College. Three years later the Center for Advanced Research began. Finally, the Center for Naval Warfare Studies created a new atmosphere for thinking about pressing national security needs for which the Navy could be used. With its gaming function, the Naval War College is a formidable PME institution. Students at the commander and captain levels study in Newport for 10 months to earn a master's degree and the Level I PME credit. The Naval War College anticipates receiving accreditation for Phase II JPME certification in fiscal year 2009.

Noncommissioned Officer Academies

Maxwell Air Force Base, Alabama
Web page: http://www.au.af.mil/au/cf/au_catalog_1999_2000/catalog2000_11_ncoa.html

The Noncommissioned Officer Academies were consolidated from ten sites around the nation to Maxwell Air Force Base to offer selected noncommissioned officers professional education to enhance their roles in the Air Force. The mission is to assist their understanding of the role that noncommissioned officers play in the national military and Air Force structures as they move toward supervisory positions.

Pinnacle

Joint Forces Command
Norfolk, Virginia
Web site: www.jfcom.mil

Administered by the U.S. Joint Forces Command of Norfolk, Virginia, the NDU oversees Pinnacle, the latest and highest level of PME in the United States. Pinnacle is a five-day course for two- and three-star officers who are going to take command of a joint task force. In the era of increasing—joint and multinational—coalition warfare, this function is viewed by many in the armed forces as a timely development. Participants have already gone through the six-week Capstone course where general officers learned much about the environment for which they have been selected. Pinnacle, however, takes things to a higher level. Pinnacle consists of a single day at Fort McNair and four days at the Joint Warfighting Center in the Tidewater area of southeastern Virginia. Upon completion of Pinnacle, general officers have a better understanding of the national policy and its implications in a particular context.

Reserve Component National Security Course

George C. Marshall Hall
Fort Lesley J. McNair
Washington, DC 20319
Web site: http://extranet.ndu.edu/rcnsc

The Reserve Component National Security Course is a two-week refresher course for reserve and National Guard officers, held twice annually in Washington, DC, as a function of the NDU. In past years, the courses were held across the country at locations such as Pensacola Naval Base and Maxwell Air Force Base in Alabama. Roughly five hundred reserve and National Guard officers convene for lectures and seminars on current regional, functional, and strategic issues. A small number of foreign officers also attend occasionally. The lecturers are the faculty at the National War College and Industrial College of the Armed Forces, the senior schools at the NDU, along with senior policymakers within the Defense Department and the NDU leadership. Participants in the Reserve Component courses also travel to Capitol Hill and other offices in the nation's capital to better sense how the Reserve Component is fitting into the national military and strategy developments.

Squadron Officer College

Maxwell Air Force Base, Alabama
Web site: soc.maxwell.af.mil

Located at Maxwell Air Force Base in Alabama, the Squadron Officer College is the newest college within the Air Force PME system and contains the Air and Space Basic Course for new lieutenants and the Squadron Officer School for captains. Both are aimed at team building, educating young men and women about leadership and airpower doctrine, and generally enhancing the careers of the next generation of Air Force officers.

Squadron Officer School

Maxwell Air Force Base, Alabama
Web site: sos.maxwell.af.mil

The Squadron Officer School, a part of the Squadron Officer College which is a component of the Air University in Montgomery, Alabama, educates captains about the culture and challenges of expeditionary warfare and airpower questions for the today's environment. Five classes of 450 students rotate through each academic year. The credit earned is applicable to master's degree programs elsewhere. The School has been a component of the Air University since 1959, although its history dates back further than that.

U.S. Air Force Senior Noncommissioned Officer Academy

Maxwell Air Force Base, Alabama
Web site: www. maxwell.af.mil/au/cepme/sncoa/

The Air Force Senior Noncommissioned Officer Academy, incorporated within the Air University at Maxwell Air Force Base in Alabama, is the institution entrusted with graduating more than 1800 master sergeant selects, senior master sergeants, selected master sergeants, and others chosen to attend. Their education here is in better skills with which to lead the enlisted personnel with whom they serve.

Notes

1. *Officer Professional Military Education Policy*, December 22, 2005, Appendix A, Enclosure A, pp. A-A-2–A-A-8.

2. Sincere thanks to CAPT Steve Camacho, USN, for discussion on this point.

CHAPTER 3

The Evolution of Military Reform and Professional Military Education

Professional military education (PME) is not a goal unto itself but both a by-product of needs the nation's military develops and an input into the development process that makes men and women into an effective fighting force rather than a series of individuals in some organization. The type of education required must be commensurate with the national military a nation, in this case the United States, has decided to develop. Because PME is the product of the changes to the system, this chapter explores some of the major military transitions in the U.S. experience.

The process of reform can be painstakingly slow or lightning quick. Yet, the evidence indicates that the debate around a major change to the military or educational system tends to percolate for a relatively long period of time, regardless of the actual passage of formal laws. Part of the reason that the debate can be so painstaking is that anything encompassed by this subject matter can have life or death effects and is not to be changed without understanding that reality.

By nature, there is a relatively tidy chronological nature to this chapter with some glaring exceptions. Because the Active Duty Component of the military has operated quite separately from the Reserve and National Guard Components, the parallel nature of their PME experiences requires separating them to show the distinctions that have existed. The discussion is not meant to lessen the importance of neither the original professional military corps nor the militia, but each developed differently, being molded into the Total Force in the twentieth century.

Professional Versus "Citizen Soldier" Concept

Long the crux of national discussion about the military role in U.S. society, still true today, is the question of whether the professional soldier ought to constitute the armed forces of the United States or whether the strongly held view of the seventeenth century would protect the nation more effectively. This latter view was that a standing professional force could become the tool of a nefarious ruler, thus the citizens would be best served by their sons and neighbors serving in the military. This would have several major advantages over a professional force, according to advocates. Citizen-soldiers could be the farmers and small businessmen that were their traditional occupations, only to don uniforms when the nation needed them to resolve a military conflict, then return to their normal lives. This would theoretically lead to fewer conflicts since being under arms would not serve as the "norm" of behavior. This would also be a less expensive option than a standing force since the bulk of the citizen-soldier's attention (and expenses) would go to his traditional occupation. Finally, this solution would offer the rulers fewer opportunities to call up their armed forces since it would require greater effort to get the forces ready to go.

Part of this debate was the unstated need for standing military education. Warfare during the period when this debate transpired was certainly simpler than today and the United States did not yet have any standing military academies, formally or informally educating the force. If a standing force were created as the method of developing a military, then institutions with set curricula were going to be required. The decision to use citizen-soldiers would have made the education the responsibility of the individuals themselves.

The first major individual to raise the level of attention to this issue was probably Baron Friedrich Wilhelm Augustus von Steuben who brought Prussian and Russian military techniques to the new nation. Von Steuben's service with the independence Army in the late 1770s and early 1780s is frequently cited as the basis to the steps to create a standing army as well as the Revolutionary Army itself.

General George Washington receives the lion's share of attention in the creation of the military but this is due more to Washington's organizational skills than his military training. Washington's personal experience was his guide to military service, based on his time as a scout in the western portion of the colonial Virginia territory and the time he spent with the Red Coat army in the French and Indian Wars of the 1760s. Washington did not have formal military education because it did not exist. While George Washington often receives credit for building a force which severed the ties with Britain and became the standing army of the United States today, he was actually a citizen-soldier in the purest sense of the concept. Between the French and Indian War (1763) and the War for Independence (1776–1783), Washington returned to his plantation at Mount Vernon, Virginia, where he was a gentleman farmer of great self-provided education and a successful business career.

Washington, for all his founding credits of the first eight years of the Republic, did not found the first PME institution of the United States. The vision for that enterprise was Thomas Jefferson's. The third president of the Republic, with his strategy for broadening and consolidating the national territory, founded an institute to develop army officers on the banks of the Hudson River in upstate New York in 1802. The eleventh president, James Knox Polk, founded the U.S. Navy School in 1845, taking on the title U.S. Naval Academy five years later.

Lessons from Global Involvement: The Root Reforms

While reforms are generally associated with more than one individual, the role of Secretary of War, then Secretary of State, Elihu Root in President Theodore Roosevelt's administration was monumental. A lawyer from upstate New York originally, Root had a highly successful career in corporate law before President Roosevelt's predecessor, William McKinley, asked him to join the administration after the turn of the century. Root entered an administration with an unconventional chief in Roosevelt and the latter allowed Root to carry out some of the most fundamental reforms ever in the U.S. government. Roosevelt, after all, was the president who wanted an army and a navy that could support his decisions to expand U.S. presence around the world. This was the first administration that had truly gone beyond the continental United States in any sustained, meaningful manner. If nothing else, the United States had acquired overseas territory as a result of victory in the Spanish-American War of 1898 so the force that Secretary of War Root sent abroad needed to be ready or to learn to address the Philippine insurgency and governance in the other newly acquired lands. The Dodge Commission, under Major General Grenville Dodge, evaluated the work that the War Department had done, leading to such strong criticism that the secretary of war resigned in 1899. The Dodge Commission focused on the logistical and health problems that were so prominent in the Spanish-American War but actually reflected deeper problems that had plagued the Army for the entire century. The Army had a commander but no chain of command as currently exists; a series of fiefdoms had developed in various parts of the Army which the War Department could not alter. The Dodge Commission made obvious the need to alter this state of affairs as the U.S. role in the world would require a more agile, disciplined Army.[1] President McKinley turned to Elihu Root. Root overhauled the entire leadership, promotion structure, education system, and approach of the U.S. Army. In the PME history, Root's reforms led to the General Staff Act of 1903 that authorized both the founding of the Army War College, under Naval War College graduate General Tasker Howard Bliss. Later, serving as secretary of state, he similarly conducted foreign policy with a shift from the status quo ante, earning himself the Nobel Peace Prize in 1912.

Education Becomes Structured

The Navy created the first Senior Service college, the Naval War College, in 1884, roughly forty years after the Naval Academy's founding in Annapolis, Maryland. During its first decades, the leadership of the Navy questioned repeatedly the value of a shore-based education that precluded the officer from being where a fine navy officer does his job: at sea. The school began to educate navy officers about their profession, allowing a systematic study of an organized curriculum. Along with this task, the Navy allowed the institution at Newport, Rhode Island, to venture into hypothetical cases where the students could test their skills. This was an early application of war gaming, a skill that is still associated with the Navy War College today. One of the most respected figures in U.S. military thought, Captain Alfred Thayer Mahan, was one of the founding instructors, gathering his lectures on sea power together in 1890 to publish them as the highly influential, *The Influence of Sea Power upon History, 1660–1783*.

As the Web site for the Navy War College admits, successive politicians sought to close the facility. World War I, with the navy successes accredited to the education gained at Newport, put the school on permanent footing and led to a greater role for the war gaming function for navy students.[2]

The Army, with its myriad of specialized training programs, created a Senior Service college at Washington Barracks in the nation's capital, beginning in 1904. Secretary of War and major supporter Elihu Root was a prominent supporter and his boss, President Theodore Roosevelt, laid the cornerstone for the building named in his honor to house the premier Army educational outfit in February 1903. The intent of the school, as true for Navy brethren, was to allow senior Army officers to study tactics, policy, strategy, and command concerns. These two institutions, since the Army soon included the Army Air Corps and the Navy included the Marines in the Sea Service component, formed the basis for the highest level PME in the country through the end of the next world war.

The Militia Evolves into the Guard

Some famous legislations passed in the past to affect the United States include the laws creating the National Guard. Originally discussed in several clauses of the Constitution, the militia were vaguely described in terms of their role and the overall congressional relationship with this armed body. The Constitutional description reflects a deep-seated political philosophy in the colonial era that questioned any concentration of arms in the hands of a government. Fear that government could oppress the citizens was widely shared and remains a fundamental political strain in the United States two and a quarter centuries later. The English experience of the Civil War in the seventeenth century, as many departed the island nation for the new colonies, left an indelible mark on the political psyche of the Founders. While Article II, Section 2 of the Constitution mentions the president as commander in chief, Article I is more detailed about Congress' role in dealing with the militias.

A decade later, the 1792 Militia Act codified a national norm for the militias. While this law contrasted with the still-strong sense of state regulation over its own assets, the decision to formulate a standard set of rules for militia in the states was a tremendous step in the evolution of the military in the nascent United States. The regulation laid out a form for militias across the states, described who was to engage in the force, and set forth what would be its role in the national defense. Prior to 1792, states had a somewhat disorderly approach to this potentially powerful tool.

The Dick Act of 1903, more than a century later, set out of the state militias under a new rubric: the National Guard. Prior to that legislation, the individual state militias, guaranteed under Articles of the Constitution and the 1792 law, were still somewhat decentralized in their composition and activities. The Dick Act allowed for a force that could be called upon as a reserve for the nation. While World War I and subsequent conflicts were far beyond the horizon at the point, the Spanish-American War of 1898, the brief experience that gave rise to the need for jurisdictional and organizational reform of the militia in the early years of the twentieth century which prepared for subsequent conflicts, produced questions about the military which reforming militia sought to address.

Lessons from World War I

The next major period of reform for the armed forces revolved around World War I. This conflict was a significant challenge to the nation because our prior armed conflicts had never had the same breadth of challenges. World War I involved the movement of a large number of forces across the Atlantic in a short period of time as the nation was still somewhat divided about the decision to engage in the conflict. Additionally, some indications were that our southern neighbor, Mexico, might pose problems which could require some more force at home. The need to balance active duty and reserve forces in a meaningful, safe way was a factor but the majority of the thought going into the World War I reforms related to mobilizing a force that could fight in the warfare which had evolved in the early twentieth century.

As true with subsequent military reform periods, the 1916 and 1920 laws, which constituted the bulk of the World War I changes, led to a change achieved over a period of national debate. The debate did not end with the return of U.S. forces in 1918 and 1919, either, but continued through a period of reflection.

The 1916 National Defense Law was a major reform of the U.S. military with several components. A major question the reform addressed was the education of military officers. Beginning in the second decade of the twentieth century, advocates for the "citizen-soldier" wanted to broaden the role of "average Joes" in the force so they sought to create educational opportunities through military camps for the elite in the United States. This idea, championed by many who were taken with the view that the threat of war required individuals to work hard at understanding what would engender peace, was known as the "Plattsburgh camps" where men would learn how to be officers without having been at West Point or Annapolis.

The nation had civilian education that fed into the military in two ways as early as the nineteenth century. Three well-known private institutions began in the early years of the 1800s to allow students possible access to the military as a career. These were Norwich College, begun in 1819 in rural Vermont, as an educational body dedicated to understanding liberal arts and sciences as well as inculcating a military discipline. Late in the following decade, the Virginia Military Institute opened its doors in 1839 as the oldest state-backed military training establishment, located in Lexington, Virginia. The third was The Citadel-The Military College of South Carolina, a Charleston body created three years later. These schools retain today the role that they first achieved in the 1800s of producing soldiers who have learned much about military affairs but are still civilians until they receive their congressional commissions as officers.

The second major access point for military education came from the Morrill Land Grant Act of 1862. The initial Morrill Act was best known for protecting mechanical and agricultural curricula at state universities. A subsidiary effect, however, was to require these colleges to offer "a proper, substantial course in military tactics complies sufficiently with the requirements as to military tactics in the act of July 2, 1862, and the other acts, even though the students at that institution are not compelled to take that course" (*Opinion of Attorney General, June 30, 1930*).[3]

This act required all federally granted schools (there are 106 today) to teach agriculture, engineering, and military tactics. Passed during the Civil War, the law prohibited universities in states at war with the government in Washington to receive the grants. When initial preparations began for what would become the World War I deployments, the Naval and Military academies, the three civilian educational colleges with strong military links, and the Morrill land-grant states became the heart of the officer corps. The need to broaden accession, however, led to a greater emphasis on education and getting those who had received or were enrolled in education at traditional universities and colleges to engage in military activities. While these "Plattsburgh camps" with their summer-long education was a notable introduction to military service, the fundamental changes that were necessary to increase the officer corps actually resulted more from World War II reform than World War I. But the need to give a grasp of armed service to a healthy side of the U.S. male population became apparent as a result of the changing U.S. role in the world seen in World War I.

The 1916 National Defense Law made it easier for men who studied under the auspices of the Morrill Act to gain commissions in the U.S. military. This had tremendous effects on the availability of officers in the frantic crush to mobilize at the beginning of the Second World War. Additionally, the changes resulting from the 1916 law led to the transformation of the military portion of the Morrill Act into the basis for the Reserve Officer Training Corps, common on many college campuses today.

The enhanced the position of the Reserves and National Guard restructured the Services to balance to the force which began the World War I military. As the 1916 National Defense Law began enumerating relationships between the active duty and reserve components of the U.S. armed services, the 1920 National Defense Amendments further integrated the National Guard into both the leadership and Army divisions.

A further effect that World War I reforms had on PME was the advent of much more specialized training and education programs. Fort Leavenworth in Kansas, established to protect the westward travelers in 1827, became the site of the Army Command and General Staff College for intermediate-level officers in 1881 under General William Tecumsah Sherman's vision. Gradually, other specific types of Army education also migrated to Leavenworth until it became arguably the central location for specialized training and education in the U.S. Army.

Education, in general, became obviously needed to prevent the reappearance of some problems that had appeared in World War I and as a stellar example of how to apply classroom thinking to practical wartime contingencies.

The Army Industrial College, built in southwest Washington, DC, at the Washington Barracks, opened its doors in 1924, partially as a result of the problems the U.S. Army faced in shipping its forces to Europe in 1917 and 1918. The Army and civilian leadership, such as the industrialist Bernard M. Baruch, realized that the nation had been ill-prepared to mobilize its massive potential force but that by studying the requirements needed to field, then deploy a large military, the United States could be far better prepared in the case of another conflict. In the early 1930s, one Major Dwight David Eisenhower both lectured and learned at the Army Industrial College, later crediting it with a tremendous impact on his ability to conduct the war in Europe. Renamed the Industrial College of the Armed Forces later, it early observed the need to broaden its appeal beyond strictly Army students to include civilians and those of the Navy.

In the more laudatory vein, the Naval War College, with its perpetual argument that line officers could benefit from schoolhouse education as much as the traditional fleet service, basked in the glory of the effectiveness of its curriculum and war games. The Navy early realized that it needed to concentrate further on educating its officers about naval tactics, strategy, policy, and command while reaching a better understanding of how this mighty force could interact with the sprawling Army, something that had proven a significant challenge in World War I.

Finally, the Marine Corps, under the leadership of Major General John Lejeune, significantly broadened educational activities for marines. Assisted by the opening of the Quantico Marine Base south of Washington, DC, the Marine Corps augmented the School of Application which had begun in 1891. After World War I, Lejeune applied education to the officer corps through the study of weapons and tactical employment at the Marine Corps Officers Training School that opened in 1919. Lejeune's successor, Brigadier General Smedley Butler opened the Field Officer's Course the following year and the Company Grade Officers Course nine months later. These three institutions became the founding blocks for the Marine Corps University which finally opened its doors in 1989 under Marine Corps Commandant Alfred M. Gray. In the decade following World War I, the Corps also started educating its officers in formal courses on Amphibious Warfare in junior and senior courses.

World War II Era Restructuring

The number of conditions that were different from World War I to World War II for the military were numerous. One major and an immediate one, however, was that officers and enlisted personnel were almost immediately in short supply. Courses during the war were briefer than usual, impacting on the "thinking" process that is now the bedrock of PME but at that time simply was not available in the early 1940s, particularly at the beginning of the war.

As a result of the World War I emphasis on education and the greater use of the young men who had studied under the portion of the 1916 National Defense Act, the nation could call upon new officers to serve in the Asian and European theaters. Without the military education these young men had had in college, it would have required significantly longer to put together the basic elements of a force to address the national needs.

Additionally, the nation's military leadership had recognized the role of some joint operations between the Army and Navy. Secretary Root had created the Joint Army-Navy Board in 1903 and it continued for sixteen years. The Naval War College had begun teaching basic joint issues in its curriculum immediately after World War I. The Army War College focused on ground issues as did the Service. During the War, military leadership in the United States recognized that these divisions into Service concerns were not serving the nation as well as a "joint" education facility would. The Joint Army-Navy Staff College moved into the Army War College building at the Washington Barracks between 1943 and 1946. Aimed at intermediate-level officers, the college worked toward better coordination between officers at the intermediate grade as the Services became more intertwined in the operations of World War II.

With the end of war in 1945, the leadership of the U.S. military began examining seriously the postwar educational reforms that would have made their recently completed tasks easier. Several immediate PME reforms appeared on the horizon. First, the Joint Army-Navy Staff College ended its tenure at Washington Barracks, opening its doors instead as the Armed Forces Staff College in 1946, first elsewhere in Washington, then in Norfolk, Virginia, where it resides today. Second, the Army Industrial College, in a bow to the need to create far more jointness across the Services, became the Industrial College of the Armed Forces. The focus of this reformed facility was to study preparation and mobilization for the nation in case of further conflicts. Its target audience became thoroughly balanced between the Services and civilians from across the national security community began to occupy important seats in this educational arena. A third action was the creation of the National War College to be an institution to prepare the future leaders of the armed services and civilian national security agencies for senior leadership in the development and implementation of national security strategy. The college was, from its inception, entirely joint (the Services had equal numbers of seats) and had a formal State Department link to the military from day one in September 1946. The National War College would replace the Joint

Army-Navy Staff College which had dislocated the Army War College which resumed its classes at a new location, Carlisle Barracks, Pennsylvania. The National War College and Industrial College of the Armed Forces became the two joint, strategic Senior Service colleges in the United States while the Armed Forces Staff College aimed at younger officers but with a look toward understanding that the future of the U.S. force was joint. The schools were not yet linked together under an umbrella called the NDU (that would occur in 1976 for the Senior Service colleges and 1979 for the Staff College) but the idea was floated.

In the 1940s, President Franklin Delano Roosevelt created a hierarchy of military leadership known as the Joint Chiefs of Staff. The predecessor organization, built by the innovative Secretary of War Elihu Root under President Theodore Roosevelt, was the Joint Army-Navy Board, an instrument of building a committee facilitating coordination in the process of decision making for war. Root had witnessed the poor coordination of inter-Service operations in a conflict as relatively simple as the Spanish-American War of 1898. Seeking to address the difficulties, the Joint Board operated through World War I. While an improvement, the United States still fumbled through the coordination steps crucial to making the various parts of the vast military work together.

World War II, however, reflected the enhanced capabilities, hence complexity, of the U.S. forces. There had been only the smallest of an Army Air Corps under the prior system along with little serious role for a fledgling Navy carrier capability as two examples. But, World War II posed a significantly different challenge in its true global nature: the two theaters for this conflict, especially the Pacific, were vast. A reorganization was absolutely essential to any serious coordination but probably even to the possibility of success.

FDR created a system of theater (the military term for the geographic space in which a military operation transpires) commanders that established the chain of command within that area. The officer with the highest responsibility in that geographic area, or AOR (area of responsibility), was the commander in chief of the specific geographic command. These jobs became known as the CINCs of the command, pronounced "sinks."

In the Services, the senior officer is the Chief. The Services each had a senior officer, a four-star general except in the Navy where the rank was four-star admiral, to whom the responsibility for administering the chain of leadership in the Service fell. For the Navy, this position was the chief of naval operations, reflecting the reality that the Marine Corps is considered under navy operations in a conflict, while the individual over the smallest Service is the commandant of the Marine Corps. The Army chief of staff is the top officer in that Service and the same would become true of the Air Force upon its formal creation with the Defense Reform Act of 1947.

To then establish a certain linkage between the theaters, the commanders were put onto a staff with the Service Chiefs of Staff. The 1940's reforms led to these Service chiefs and the combatant chiefs to serve on a single unified entity called the Joint Chiefs of Staff under the leadership of the Chairman of the Joint Chiefs of Staff. The Chairman, of JCS, is an appointed position for a four-star officer for a two-year term, renewable in most instances. The Joint Chiefs of Staff in its initial years quickly developed a staff to support the Chiefs and implement the objectives of this joint body. This latter group was the Joint Staff.

While containing the leadership of both Service and combatant commands, tensions immediately developed between the officers in the Services, those of the combatant commands, and the Joint Staff. Each group had a mission it sought to accomplish in a resource-constrained environment, thus creating an impression that other groups represented an amorphous "threat" of some sort to the mission of a single group. In the aftermath of World War II, as the nature of the threat to the nation changed, the differences between these groups' concerns and goals ossified to the point where the missions of the various groups seemed virtually in conflict to the members of each group. The incentives to improve coordination and cooperation between these various actors fell.

The captain of a ship operates in a relatively autonomous position with a set of orders to accomplish a mission; his or her job is governed by that set of circumstances far different from their hierarchical structure of the Army or the Marine Corps. The Navy simply has always feared that consolidation of control would adversely affect its ability to accomplish its mission while the other Services have been more open, in the main, to a national consolidation of force and leadership.

The 1947 National Security reforms dramatically altered the community through creating a permanent security apparatus and reformulated the leadership of the military itself. The intelligence community grew substantially with the Law's creation of the Central Intelligence Agency out of the wartime Office of Special Services. The Army Air Corps became an independent Service and ultimately developed its own education academy, the Air Force Academy in Colorado Springs, Colorado, and the Air War College and attendant schools at Maxwell Air Force Base in Alabama. The 1947 revisions were wide-reaching and fundamental to the structure of the national security community of the United States.

President Dwight David Eisenhower had a unique perspective on the circumstances that he faced in the 1905s. Having served as the combatant commander during the final years of the war, followed by his tenure as Service chief, Eisenhower had seen the operations of Navy and Army forces in the European campaigns that brought down the Nazi and Fascist regimes. He understood the difficulties of orchestrating a naval fleet to transport a massive army to a highly defended coast where a land invasion was the objective. Further, Eisenhower left his European Command position in favor of taking the position as Army chief of staff where he witnessed the goals of his staff in seeking budget and other resource allocations from a changing military.

In the last years of his presidential administration, Eisenhower sought to bring significantly better genuine coordination and legal balance to the system. His reform attempts, spotlighted in 1958, would have limited the role of the Services in this system, bringing the Navy with its tremendous independence into a more mutual operational perspective with the other Services. Even with his tremendous personal experience and the respect he earned as having "been there, done that," Eisenhower did not achieve the reforms he sought in the late 1950s. The lessons of the Services' power in the U.S. military at the height of the Cold War threats were unavoidable: powerful protectors in the Congress would thwart efforts to alter the status quo even in the face of decreasing the efficiency or increasing the cost of the national security effort for the United States.

Vietnam Era Evolution

The war in Southeast Asia, more often called Vietnam, lasted roughly from 1960 through 1975 but it was a hard one on U.S. society. A particularly important aspect for the PME community related to the changes that the war brought to the Reserve Officer Training Corps (ROTC) programs around the nation.

As the war continued in the late 1960s, particularly as student protests increased and public support waned, the environment on college campuses became less welcome to uniformed officers teaching classes that were the basis to the ROTC programs. The disapproval was far from universal across the nation but at many large universities protests did result. In the most tragic case, a bombing of the ROTC office in the Math Research building at the University of Wisconsin Madison resulted in death.

The Morrill Act of 1862 and the subsequent 1916 National Defense Law made ROTC a requirement of male students receiving aid at these colleges. The uproar resulting from the emotional protests against the War led to the downgrading of ROTC programs to a far more voluntary status. This change led to a shift that accompanied the end of the lottery and introduction of a voluntary force.

President Nixon ended the obligatory draft, opening the door to volunteer service in 1973. This led to a military based on some of the principles that had characterized the armed forces throughout their history—professional soldiers, sailors, airmen, and marines who sought to serve their nation—while allowing different forces, such as self-selection, to creep in. The military of the early twenty-first century is overwhelmingly southern, as an example, and there appears a distance between the military and some elements of society—those who are the decision makers.[4] Fewer congressmen and women today have military service than was true for the Cold War, for example. The 1973 decisions also opened the door much wider for women to participate fully in the armed forces of the United States, a radical shift from the past.

The 1973 presidential decision to end the lottery draft by which young men received their notices of required national military service was a tremendous shift for the twentieth-century military. Similarly, when President Jimmy Carter decided by executive order both to pardon the young men who had ignored their military obligations and to reimpose a system of military registration as a precursor to some unforeseen instance where a mass call up were necessary.

Goldwater-Nichols Changes

The Vietnam experience led to disheartenment in the U.S. military. Several military leaders of the First Gulf War (1990) have written memoirs which discuss how they approached rebuilding the armed forces after the war ended and the All-Volunteer military began. Drug, race, and chain-of-command issues plagued the Services and had to be addressed. In the late 1970s, these factors led many to ask whether there were ways the defense spending should go to other programs, including taking out of PME. It was a bad period for defense with a growing sense something had to be done.[5] PME was a method of doing so but its role was actually more important later.

Occurring roughly the same time as some of these concerns, the Clements Committee on Excellence in Education, named after Deputy Secretary of Defense William Clements, provided outside suggestions on curriculum development (consolidation of a unified curriculum at ICAF and NWC) and academic ambiance. Noteworthy recommendations, buttressed by the Skelton reforms of the late 1980s, included "not-for-attribution" and open, seminar "peer" discussions. The intent, as true with Skelton later, was to raise the quality of the education the taxpayer invested in. The result—the NDU—appeared as a reform in 1976. It brought together the Industrial College of the Armed Forces and National War College under a single umbrella.

The problems confronting the military included the failure or near-failure of a number of operations of the U.S. forces. The *Mayaguez* incident in the final days of the Vietnam War, 1975, was one such embarrassing difficulty when Khmer Rouge forces from a Cambodian island pirated the S.S. *Mayaguez*, holding the crew briefly. More significant was the utter failure of the *Desert One* rescue mission for forty-four hostages in Teheran, Iran. Rather than whisk the long-held U.S. officials from the hands of Iranian extremists, death and destruction resulted from poor communications and coordination between aircraft and personnel in the desert.

In the early 1980s, other examples of poor coordination and communication plagued the U.S. military. The October 1983 Marine Barracks bombing of Beirut, Lebanon, occurred, to some extent, because communication and force protection difficulties made adequate assessments of the requirements to protect forces fail.[6] There was no rationale for the terrorist to drive a car into the building early one Sunday morning but after action reports indicated that poor organization within the U.S. forces left them vulnerable.

At the same time, the joint operation in Grenada, also in late October 1983, exposed the weakness of the term "joint." Units on the small Caribbean island found themselves unable to communicate with the other Services because the equipment lacked standardization. Transportation across the Services proved a surprising challenge. Questions from studies after that strongly criticized the Services for their unwillingness to look for collaborative systems, protocols, and such.

The accumulation of instances led to congressional and executive branch pressures for reform of the armed forces. Similarly, groups within both branches resisted reform strongly.[7] The half decade struggle had two major supporters who gave gravitas to the effort: The chairman of the Joint Chiefs of Staff, Air Force General David Jones (1978–1982), who publicly testified in 1982 to the need for this reform (before the 1983 incidents which reinforced his warnings), and Arizona Republican Senator Barry M. Goldwater, who was a general in the Air Force Reserves. Other supporters instrumental in the process included Georgia Democratic Senator Sam Nunn and a long list of professional staffers.

The intent of Goldwater-Nichols reforms was to create a structure which would better serve the nation. The role of the chairman of the Joint Chiefs, in place for forty years as one of the heads of Services or commands, became the primary advisor to the president of the United States. The goal was to undercut the bureaucratic strife resulting from Service chiefs protecting their turf rather than considering the overall defense posture. The commanders in chief of the regional and functions commands, as they were designated from the 1940s until Secretary of Defense Rumsfeld renamed them combatant commanders, were to employ the range of Services in that command instead of favoring on his own command.

Education took on a significant role in fostering jointness by requiring the Navy, traditionally less sold on the importance of structured, schoolroom thinking, to allow its officers to pursue education and fulfill the same joint education requirements as the other Services. A portion of the Goldwater-Nichols legislation made joint professional military education (JPME) certified at the Phase II, advanced, level that would put Navy officers on the same footing as their cohorts from the other Services. Traditionally, Navy officers earned their promotions from being line officers deployed at sea, whether in a submarine, on a surface vessel, or as a naval aviator. The duty at sea has been the measure of learning and success, leaving little time for shore duties or formal education. Air Force, Army, and Navy officers, however, had the option of pursuing PME or civilian education such as master's or doctoral degrees. The pursuit of advanced degrees often put these officers in a more advantageous position for promotion. Navy officers who did earn advanced degrees often did so when deployed as a ROTC instructor or through the Naval Postgraduate School in Monterey, California. These choices, however, were still time away from the fleet.

The role of education in assignments was also important to the revisions. Prior to the 1986 Law, critics noted that the quality and experience of officers assigned to the Joint Staff was lower than of those assigned to their Service billets. The Services, it appeared, did not respect the joint duty or experience while expecting a much higher level of specialization of its forces.[8] Because an officer was expected to excel at his Service specialty, few officers sought to attend the purely "joint" programs such as the Armed Forces Staff College, National War College or the Industrial College. Statistics indicated that a surprisingly low number of officers had studied at the schools which preached joint efforts. A goal of the legal revision in 1986 was to increase dramatically those numbers.

From Goldwater-Nichols came Joint Specialty Officers (JSOs) that are essential to carrying out a better system of jointness. JSOs, under Title IV of the Law, meet established criteria and address stated curriculum requirements to better the pool of officers who will serve in joint billets. Those billets are established in a list called the Joint Duty Assignment List which denotes the specific intent of Congress to increase the desirability of these assignments. This was often, particularly in the years immediately after the Law took hold, viewed as both detrimental to the fabric of the individual Services (as each has its own desired curricula foci) and too high a number relative to the Services' needs. Goldwater-Nichols and its defenders, however, have monitored the JPME curruicula through periodic reaccreditation by the Joint Staff office, which oversees education, the J-7: Joint Plans and Interoperability. The bottom line for jointness and education is that a minimum of 50 percent plus one of the graduates from JPME institutions (War Colleges and Joint Forces Staff College) must take a JSO-listed job immediately upon completing the education, for a stated minimum period. JSOs have played a tremendous role in creating a truly joint staff.

Along with the PME requirements that the legislation instituted, two other critical elements exist to stimulate the "joint" nature of the military as Congress desired. First, an officer seeking flag or general officer rank must serve in a "Joint Specialty Officer" billet for a minimum of two years' duration. These positions are specifically coded as joint by the Joint Staff to provide the incentives to have officers accept the assignments, a condition not true prior to the reforms. Second, the Services must promote their officers who serve in JSO billets at the same rate as they promote individuals who remain in Service-based jobs. This requirement helped draw more highly qualified officers to the Joint Staff positions rather than seeing them reject such positions for fear they would jeopardize career-advancements, as had occurred historically.

Goldwater-Nichols, and subsequent congressionally mandated revisions to the PME program, created this state of affairs. After 1986, with "grandfathering" for those already senior enough in uniform, all officers who achieve flag or general officer rank must have met the Phase II JPME certification. The Staff finally had a role in PME, not just JPME, which had never been the case before.[9] The Military Education Coordinating Committee, composed of all applicable Defense Department institutions in education, oversaw this function. Joint education was the overall goal.

Further, largely at the behest of Missouri Democratic Representative Ike Skelton, the Senior Service colleges (Army, Navy, Marine, Air Force, Industrial College of the Armed Forces, and National War College) have all pursued formal academic accreditation for their ten-month programs. Each of these schools now have certification from the relevant accreditation body (in the case of the National War College, the Industrial College, and the Army War College the institution is the Middle States Commission on Higher Education in Philadelphia, Pennsylvania).

At the same time, the Industrial College mission grew to include study of acquisition by the military community. The Defense Acquisition Workforce Improvement Act of 1990 was an outgrowth of the Packard Commission of 1985 that had evaluated the quality of defense management. The 1990 Act was to guarantee a much better quality to the workforce to allow DoD to more effectively and efficiently conduct acquisition activities and broadening its personnel understanding. The Defense Acquisition University became a major subfunction of the Industrial College as it homed in on providing this education to the military community.[10] This was an indication of the way that the NDU could help accomplish broader goals of the U.S. government and the Defense Department over national security issues alone.

CAPSTONE and Pinnacle

Along with the development of the JSO requirements, Title IV of the 1986 legislation requires that newly promoted (either named on the promotion list or actually pinned on with the rank) flag and general officers take a six-week course at the NDU called CAPSTONE. Initially a voluntary program created in 1982, Goldwater-Nichols refocused on this non-degree-granting program to alert new senior officers to the role of jointness in their future work. The course allows new brigadier generals, rear admirals (lower half), and newly selected ambassadors to interact with retired four-star officers who participate, by invitation, with the group for the duration as Senior Fellows. These Fellows have served at the highest levels of the armed forces so they can point out the pitfalls and opportunities ahead. The course consists of travel overseas and domestically, case studies, exercises, and seminars on national security with an eye to jointness. The course introduces participants to the concerns of the various combatant commands and joint issues which cross commands.

A more recent development is the Pinnacle program, also conducted by the NDU at the Joint Forces Command in the tidewater area of southeast Virginia. Pinnacle is targeted more directly at combined and joint task force commanders who will work with joint and multinational efforts around the world. The bulk of the officers in the Pinnacle course are three-star officers, although a separate, similar five-day course is open to two-star commanders. A parallel, five-day course labeled Keystone is available to senior enlisted personnel in their command capacity. The purpose of Pinnacle and Keystone courses is to give U.S. military personnel the best preparation for the changing environment in which they find themselves as global commitments change.

Post-9/11 Pmeas Radically Different or Considerable Continuity?

Military reform appears on a periodic, if not cyclical, basis in the United States. Institutional interests prefer to keep the status quo ante to preserve equities if not perquisites and advantages, financial and otherwise. These interests do not do so out of personal greed or institutional aggrandizement; they do so, in the overwhelmingly large number of instances, out of true belief in the power of the existing system.

Military reform is also necessary to maintain the innovative thinking, or "cutting edge" advantages, which propel an armed force to success or relegate them to deadly failure. The Goldwater-Nichols Military Reform Act of 1986, hammered out over the better part of the decade of its passage, exemplified this effort and remains the last formal legislative attempt to alter the Services. It is by no means bound to be the last for the United States.

For much of the past twenty years since Goldwater-Nichols was passed, those evaluating the progress of the armed services have felt that significant improvements have occurred. This is not to argue that tensions between the Services and the Joint Staff, for example, do not persist as they do. The lessons of the wars in Iraq and Afghanistan are constantly under study and revisions to the PME program are inevitable.

The current overall concern in many circles is that the civilian ties with the defense community need clarification and improvement. This frequently leads to whether some sort of fundamental reshaping of the interagency process is imperative as the United States becomes more heavily involved in global operations of a joint, coalition, and/or multinational nature.

A "Goldwater-Nichols" for the InterAgency?

September 11 was the Pearl Harbor for the current generation. Both events were wake up calls for the national security community to reconsider not only its preparation for conflict but the level of communication, cooperation, intelligence sharing, and other aspects of conducting agencies conducting their business. Whether better coordination could have prevented either 1941 or 2001 will always remain an open question but addressing the effects would certainly be easier if agencies knew who had what capabilities at any particular time.

One of the most common suggestions for enhancing homeland security in the first decade of the twenty-first century is the passing of a "Goldwater-Nichols-type" reform package for the array of agencies in homeland security.

There have been a range of suggestions and studies on the question. Chairman of the Joint Chiefs of Staff, Marine General Peter Pace, advocated this sort of reform in a 2004 speech, while still serving as the vice chairman of the Joint Chiefs.[11] One of the nation's most respected think tanks, the Center for Strategic and International Studies, launched a three-part investigation of the question, producing a Phase I report in March 2004, to find practical help that will provide some immediate answers. The two subsequent portions of the study have investigated current government and military command structures as well as the likely problem areas ahead. The Phase II report appeared in 2005.[12] The final portion of their study will discuss command structures, role and integration of the Reserve Component, and other concerns, found at the CSIS site at http://www.csis.org/isp/bgn/.

Those directly involved in the process have also proven warm to the idea that reform within the interagency and across agencies cannot be ignored. War College curricula, such as that at the Industrial College and National War College, have asked their students to wrestle with the issues.[13]

The need for modification or restructuring, however, does not equate to conducting it in the same manner as the military reforms of Goldwater-Nichols. For one thing, civilian agencies tend to be far more decentralized than are even the military services with their chains of command and hierarchical structures. Additionally, as the creation of the Department of Homeland Security illustrated (discussed below), there are definite advantages and *unavoidable trade-offs* to amending any governmental structure. This is not to argue that the costs, fiscally and metaphorically, are too high to pay but too often advocates of reform discuss the necessary changes as if the costs did not exist. The Founders of the U.S. political system intended it to be a slow, deliberate process even if that strikes those of us in the twenty-first century as distinctly anachronistic. Completely open to question, however, is how difficult this reformulating the interagency process into something akin to the defense process would be to achieve.

How Does Homeland Security Mesh with National Security?

One of the major effects of the changes implemented by the government after the September 11 attacks was a realignment of agencies under the Department of Homeland Security. The lessons from these reforms are not unambiguous successes, however, and bear consideration in looking toward future reforms.

The movement to increase the safety and security of the continental United States, euphemistically referred to as the homeland, did not begin on September 11, 2001. With the attacks on various U.S. military units abroad and attacks on leadership of other countries, voices advocated creation of a greater protection for the homeland in the 1990s. The Senate began considering homeland defense issues in the latter part of the decade. Much debate then revolved around whether the appropriate term was homeland defense, national defense, or homeland security.

With President George W. Bush's administration, homeland security was an advisory position. President Bush had opposed suggestions that he create another federal cabinet-level office. Former Pennsylvania governor Tom Ridge did not serve as Homeland Security advisor until October 2001. The Homeland Security Department itself did not emerge until 2002. At that point, the vast shift of the bureaucratic entities that relate to homeland issues occurred. This was one of the most comprehensive movements of agencies that the U.S. government has ever seen, with questions remaining four years later as to the wisdom of some of the transformation.

The Homeland Security Department established Directorates. Border and Transportation Security absorbed the U.S. Customs Service (from the Treasury Department), the Immigration and Naturalization Service (Justice), the Federal Protective Service, Transportation Service Administration (Transportation), Federal Law Enforcement Training Center (Treasury), Animal and Plant Inspection Program (Agriculture), and the Office of Domestic Preparedness (Justice). The Emergency Preparedness and Response Directorate folded in the Nuclear Incident Response Team (from the Energy Department), the Federal Emergency Management Agency (formerly independent), the Strategic National Stockpile and the National Disaster Medical System (Health and Human Services), National Domestic Preparedness Agency (FBI) and Domestic Emergency Support Teams (Justice). For the Directorate of Science and Technology, four "assets" from various portions of other agencies were transferred. In Information Analysis and Infrastructure Protection, four agencies moved into the Directorate from other portions of the government. Two large components, the Coast Guard (from the Transportation Department) and the Secret Service (from Treasury) also switched. The sum of these changes was to move wholesale functions from one department to another but not necessarily because the new location would be any better. As response to Hurricane Katrina indicated in August and September 2005, the Homeland Security functions are far from adequately integrated to achieve the goal of seamless response to crisis in the United States.

Similar effects would seem true for educational activities. While PME relies on a relatively circumscribed field of study, homeland security—a related field—is somewhat more amorphous and still evolving with shocking speed.

The College of National Security and National Security University

The need for greater coordination than just provided by the 1986 law became immediately apparent after 9/11. Within a couple of years, that topic wound its way into the governmental educational structure. There was, however, no focus or even an easily identifiable location for creation of a curriculum, a faculty, or a schoolhouse. What many people did agree upon, however, was that the Department of Defense provided a model through its NDU.

In 2003, former assistant secretary of defense Steve Duncan began serving the NDU president as principal advisor on Homeland Security concerns. The following year, Lieutenant General Michael Dunn of NDU announced the formation of an institute for Homeland Security Studies which would develop a curriculum and coordinate coursework for military and civilians in this area. As the Institute's Web site, http://www.ndu.edu/IHSS/index.cfm?pageID=100&type=page, indicates, the Institute has links to courses offered at the National War College, Industrial College of the Armed Forces, the Joint Forces Staff College, the Information Resources Management College, and the smaller components of the NDU. The Institute's Director has been engaged in publishing op-eds and articles that illustrate the concerns this challenge offers to the nation and the national security community.

The 2006 *Quadrennial Defense Review* makes reference to the creation of a National Security University. While this document enunciates goals for the U.S. government at a high level of analysis that does not include implementation trade offs, discussions about the need for a transformation of the professional military institution have been going on for years.

The advantages of reconfiguring the NDU into a National Security institution include being current with the issues most of interest to the White House and the leadership of the Defense Department. The pressures within the ranks of uniform officers and civilians to be prepared for other attacks on the homeland while learning the mistakes that allowed the 9/11 attacks to occur are high. Finding a location in the United States and specifically within the Defense Department to study these concerns is highly advocated by many. Other supporters argue that the only way to break the stranglehold of some traditional security concerns, such as Secretary of Defense Donald Rumsfeld has tried doing through armed forces transformation, is to begin putting substantial investment in homeland security education.

At the same time, others are concerned that the issues posed in homeland security are not really new at all. This point of view argues that what occurred on 9/11 was not all that different from other security threats to the nation and must be seen in that long-term view of national security. Scholars and practitioners with this perspective argue that radically changing the orientation of security toward homeland security runs two significant risks of its own. First, the emphasis on homeland security would divert attention from security in general, undermining preparedness for subsequent attacks. Second, diverting resources and strategic thought to homeland security takes away both the awareness and the defense strength that has protected the United States against traditional and highly dangerous external threats to U.S. interests. People of this view tend to see homeland security issues as merely a different manifestation of traditional balance of power politics to which the security community will develop a meaningful, appropriate response.

Notes

1. Terrence J. Gough, "The Root Reforms and Command," at www.army.mil/CMH-pg/documents/1901/Root-Cmd.htm.

2. http://www.nwc.navy.mil/L1/History.htm

3. http://www.nasulgc.org/publications/Land_Grant/LandMorrill.htm

4. Richard Kohn and Peter Feaver, *Soldiers and Civilians: The Civil-Military Gap and American National Security* (Cambridge, MA: MIT Press, 2001) is a study based on extensive polling data from uniformed personnel and civilians. While one analysis of U.S. society that

is far from universally accepted, Kohn and Feaver's analysis is often cited on a gap in U.S. society. The data for this study predated the 911 attacks and the Iraq war.

5. John Yaeger discusses these growing problems and how they affected the Services and overall national security in *Congressional Influence on National Defense University* (unpublished dissertation, George Washington University, 2005), pp. 75–80.

6. Gordon Nathaniel Lederman, *Reorganizing the Joint Chiefs of Staff: The Goldwater-Nichols Act of 1986* (Westport, CT: Greenwood Books, 1999), p. 65.

7. For the best description of the process by a major participant, see James Locher, *Victory on the Potomac* (College Station, TX: Texas A & M University Press, 2003). Locher was a Senate staffer with an inside view of the entire process resulting in the Goldwater-Nichols reform of 1986.

8. Lederman, *Reorganizing the Joint Chiefs of Staff*, p. 43.

9. Yaeger, *Congressional Influence*, p. 146.

10. Yaeger, *Congressional Influence*, pp. 221–224.

11. Jim Garamone, "Pace Proposes Interagency Goldwater-Nichols Act," *DefenseLINK*, September 7, 2004.

12. Clark A. Murdoch and Michèle Flournoy, eds., *Beyond Goldwater Nichols: U.S. Government and Defense Reform for a New Strategic Era Phase 2* (Washington, DC: Center for Strategic and International Studies, 2005).

13. An award winning essay from 2004 was written by Martin Gorman and Alexander Krongard, "A Goldwater-Nichols Act for the U.S. Government: Institutionalizing the Inter-Agency Process," *Joint Force Quarterly*, 39: 51–58.

CHAPTER 4

Personalities

Baruch, Bernard Mannes (1870–1965)

Bernard Baruch was one of the most influential industrialists of the twentieth century, leaving an immeasurable mark on logistics and more broadly professional assessment of warfighting in the United States. Born in Camden, New Jersey, to a Civil War surgeon, Baruch graduated from the City College of New York in 1889 when he went to work by running office tasks along Wall Street where he aspired to become a participant in more significant work. Gradually, Baruch rose through the ranks at Wall Street firms, gaining much success as a broker on the Street, until he and his brother, in Baruch Brothers, were captains of the investment industry during the first decade of the twentieth century. President Woodrow Wilson called upon Bernard Baruch to help the United States mobilize its woefully inept military for conflict in Europe. Baruch served on the War Industries Board and assisted President Wilson at the Paris peace negotiations in 1919. Baruch, a millionaire several times over, continued to serve the White House, regardless of the party in power, through President Truman after World War II. Baruch was instrumental in the creation of the Industrial College of the Armed Forces, an educational institution that has the focus for the nation on the resources and strategy necessary to achieve success in promoting national security issues around the world. Baruch's contribution is memorialized at the Industrial College, which is located in Eisenhower Hall, in the Baruch Auditorium, the location for combined student convocation at the NDU.

Bliss, Tasker Howard (1853–1930)

An Army lieutenant in the initial class at the Naval War College in Newport, Rhode Island, in 1884, Tasker Bliss learned so much that he went on to become the first president at the Army War College at its founding two decades later. Born in Pennsylvania, Bliss attended West Point before receiving his commission. His early years in service were spent teaching at West Point, although he eventually went to the Naval War College where he taught. He brought new ideal to the Army.

Clark, Grenville (1882–1967)

Grenville Clark was a crucial individual in the earliest moves toward professional military education (PME) in the United States, although his effects originally came from educating civilians during the period prior to World War I. Clark was a wealthy New York lawyer who circulated in the same elite circles as many government officials in the first decade of the twentieth century. When World War I broke out in Europe, Clark began working with others concerned about U.S. preparation should the conflict drag in the United States. He also worked on bringing young men of his upper class status into military preparation, an awkward need in U.S. society where military service was limited and certainly not a career-enhancing activity. Clark was one of the major supporters of the civilian education camps that took place in upstate New York and other northeast locations in 1915–1917. With his Dartmouth background, Clark did not advocate these men going to the Service academies but believed they needed training to prepare them to accept military commissions should the need arise. The civilian camps were largely overtaken by President Wilson's decision to enter the conflict in 1917. Clark remained active in the national debate on security through the remainder of his life, with his most famous writing being the 1958 *World Peace Through World Law*. Clark remains associated with the Plattsburgh Movement.

Cohen, William (1940–)

This former Maine Republican senator served as President Bill Clinton's third secretary of defense, after former congressman Les Aspin of Wisconsin and Dr. William Perry of Stanford University. Senator Cohen crossed the political aisle to assume this leadership role much as he proved willing to work with partisans of both sides throughout his congressional career. Cohen had been a lawyer and mayor of Bangor, Maine, when first elected to the House of Representatives in 1972. Following three terms in that body, Cohen sought and won election to the Senate as the junior senator from his state. He served on important committees in the Congress before President Clinton chose Cohen to serve as the defense secretary during his second term. Cohen's position as secretary of defense overlapped with U.S. involvement in NATO-sanctioned conflict in Kosovo as well as a range of other peacekeeping efforts around the world in places like Haiti. Secretary Cohen oversaw an expansion of joint professional military education (JPME) both to include civilians in understanding the issues and bring in international officers and civilians through outreach that was intended to assist PME for U.S. officers. First, the Defense Leadership Management Program (DLAMP) began at

Ft. McNair in Washington, DC. This was an educational course to give Department of Defense civilians an adequate background in the issues confronting their military colleagues and to educate these civilians about the issues involved in the development of national security strategy. A portion of the initial classes were sent to the National War College and the Industrial College of the Armed Forces, the two Senior Service colleges which have always had a "joint" (all Services and the State Department) student and faculty balance. The majority of the people selected for each DLAMP class (lasting roughly three months) went into a small program that has evolved into the School for National Security Executive Education (SNSEE), which has taken on an increasing role in counter-terrorism training. The SNSEE emphasis, however, remains on providing access to U.S. government civilians about the national security issues confronting the United States. During Cohen's tenure, another broad expansion of PME occurred. The NATO-controlled Marshall Center in Garmisch, Germany, and the Asia Pacific Center for Security Studies, under the reigns of the Pacific Command, in Honolulu became firmly established to interact with foreign officers. In the case of the Marshall Center, the intent was to help the states of the former Soviet empire develop their cadres of knowledgeable individuals in the fields of civil-military relations and democratization of the military process. The Asia Pacific Center had a slightly different task but it operated in the same paradigm to educate the new leadership of states turning to democratic governments in East Asia. The Asia Pacific Center was as much a site for interaction between representatives of a vast region of the world. The curriculum has regional components as well as functional issues such as transnational concerns, energy issues, and militarization of the region. Later in Cohen's time in office, the Center for Hemispheric Defense Studies and the Center for Near East and Southest Asia were established in 1997 and 1998, respectively, at Fort Lesley J. McNair in Washington, DC, even though the Centers were not under the control of the NDU at the location. The Center for Africa Security Studies had its headquarters at Ft. McNair but was the only regional studies center which actually moved its seminars around the African continent to better meet the needs of the militaries and societies it served. All of the Centers have evolved in their work and have been under the Defense Security Cooperation Agency since 2005. Secretary Cohen left office in 2001 with the advent of the George W. Bush administration, with his postgovernment work in private consulting.

DeWitt, John Lesense (1880–1962)

Lieutenant General John DeWitt was a member of the generation born in 1880, along with Generals George Marshall and Douglas MacArthur, who became the architects of the victory in World War II's global theater. A Nebraskan by birth, DeWitt received his commission as an infantry second lieutenant in 1898 before assignments in the newly acquired Philippines and World War I time in Europe.

In the early 1930s, DeWitt became the quartermaster of the Army. Brigadier General DeWitt became major general when he was commandant of the Army War College in southwest Washington, DC, between 1937 and 1939. In this role, DeWitt was instrumental in creating the Army educational program that officers took to the battles of World War II since students at the War College were at the senior ranks of their Service as they studied. From the War College leadership job, Major General DeWitt pinned on a third star in 1939, assuming command over the Ninth Corps and Fourth Army at the Presidio in San Francisco. His tasks included the training of Army officers, ground and air components, in the western portion of the nation, meaning thousands of men on active and reserve status were under his leadership. DeWitt engineered exercises and maneuvers for these forces while upgrading the quality of the installations along the west coast to include the Presidio in San Francisco. DeWitt worked on joint maneuvers, some of the first joint military operations, between some of his Army units with Navy forces along the Pacific coast.[1] As commander of the Western Defense Command after December 7, 1941, Lieutenant General DeWitt carried out one of the more infamous tasks in U.S. history by interning Japanese-Americans. The nation questioned the loyalty of these citizens and DeWitt executed the order to put them into containment camps for the duration of the conflict.[2] Lieutenant General DeWitt became head of the Joint Army-Navy Staff College between 1943 and 1947 when he retired from active duty. In that position, DeWitt's views on JPME, a topic of considerable attention as the post-World War II defense education architecture was being considered, were highly sought. Promoted to full general after his retirement, DeWitt died in 1962.

Eisenhower, Dwight David (1890–1969)

General Dwight David Eisenhower had an entire career in uniform, replete with tremendous success, prior to assuming the presidency of the nation in January 1953. Eisenhower, born in north Texas, moved to Abilene, Kansas, where he spent most of his childhood as the third of seven sons. A phenomenal athlete, Eisenhower applied for and secured an appointment to the Military Academy in West Point, New York. Not a particularly strong performer at the Academy, Eisenhower graduated in the upper half of the Class of 1915, known as the Class that "Stars Fell Upon." During his first assignment as a second lieutenant at Fort Sam Houston in Texas, Eisenhower met Mamie Doud to whom he was married for sixty-three years until his death. Eisenhower's career between graduating from West Point in 1915 and the early 1940s was a relatively uncommon one because he excelled as the Army drew back down in size. He served in the Canal Zone where he became a student of history, strategy, and critical thinking, which prepared him for his assignment at the Command and General Staff College (CGSC) in Leavenworth, Kansas. Eisenhower had improved his academic prowess dramatically by graduating first in his CGSC class, earning considerable attention by those in the Army that he had an excellent mind that could be applied to pending problems. General Jack "Black Jack" Pershing appointed Eisenhower to the American Monuments Commission, giving him access to Europe and Washington, DC. From there, Eisenhower had a major role, underappreciated, in the creation of PME, however, by serving as one of Army planners for mobilization. In that position, Eisenhower examined something crucial to carrying out a sustained, complex conflict should it ever arise. He began lecturing at the recently (1924) created Army Industrial College in 1931, highlighting the need for thinking on how to provide resources in the long supply lines that would result if a war occurred again in Europe or, worse, in the Pacific. From this Army staff position, Eisenhower became an aide to General Douglas MacArthur, accompanying the General to Philippines in the late 1930s with the assignment of creating a Philippine military in preparation for the islands' independence in 1946. Generally seen as a backwater from the Army, General George C. Marshall requested Eisenhower return to the United States in the months immediately prior to the Pearl Harbor attack. In various short assignments back stateside, Eisenhower proved exceptionally good at applying the tasks he had studied over the prior decade. Marshall, then Roosevelt's chief of staff of the Army, chose Eisenhower for increasingly important positions such as War Plans for the Army. In 1942, Eisenhower was made Major General, meaning a move to build coalitions in Europe with the Allies. Late that year, Eisenhower became combined commander for the forces which would carry out *Operation Torch* in 1943. Later that year, Eisenhower took on the job of the vast Normandy invasion known as D-Day as the Supreme Allied Commander in Europe. In 1945, he served in that position until the Axis' surrender in May when he became Military Governor in Europe. Upon his highly feted return home, Eisenhower became chief of staff of the Army. In 1946, he and General George Marshall, along with other top leaders from the recently completed conflict, worked to create the National War College, another senior PME institution. The War College, unlike its counterpart the Industrial College of the Armed Forces where Eisenhower had taught in the early 1930s, was to focus entirely on the study of national security strategy. The institution was to be entirely "joint," meaning that equal parts of each student class would come from the Sea Services (Navy and Marine Corps), Army, and (when created in 1947) Air Force, along with a standing participation by the Department of State. This unique institution would hopefully lead to better understanding across and within the Services as well as across the government as a whole to prevent the "friendly fire" incidents that had killed Lieutenant General Lesley J. McNair behind allied lines in France in 1944. Upon retiring from active duty, he assumed the presidency of Columbia University. In 1952 he defeated Illinois Senator Adlai Stevenson for the presidency, a feat he repeated four years later. Eisenhower retired from the presidency to his farm in southern Pennsylvania where he died in March 1969. Eisenhower's role in professionalizing the military in the United States is difficult to overstate.

Flipper, Henry Ossian (1856–1940)

The initial African-American graduate of the U.S. Military Academy at West Point, Henry Ossian Flipper was a pathbreaker in U.S. PME. Born into slavery in Georgia, Flipper received an appointment to the Academy in 1873 and graduated four years later. His first assignment as a commissioned officer, between 1878 and 1880, was to the southwest of the United States where he had the traditional Army position as a surveyor. The lieutenant was charged with embezzlement and "conduct unbecoming an officer," resulting in his departure from the Army in 1882. Flipper went on to serve as a government official and surveyor of note in civilian life. He also became a noted author, publishing several volumes on a range of topics to include Mexican law on mining and autobiographical works on his experience. He died in 1940 and over the next fifty-nine years, his descendants used his lifelong declarations of innocence to have his record of "behavior unbecoming an officer" expunged from his professional record. President Bill Clinton pardoned him and noted the wrong that had been done to Lieutenant Flipper.

Gerow, Leonard Townsend (1888–1972)

The commander of the Fifth Corps that saw heavy action in World War II on June 6, 1944 at the Normandy invasion, Lieutenant General Leonard T. Gerow was an experienced leader who had been first in his class at the Virginia Military Institute (1911). Gerow was one of the elite officers from the Army who reached the zenith of their careers as World War II began, serving as the chief of the Army War Plans Division in late 1941. Gerow held senior Army positions before being the commander of the Fifth Corps and then, with its mobilization in late 1944, the Fifth Army in the last months of the war. Toward the end of World War II, Gerow became the commandant of the Army CGSC. He then held a crucial position in the World War II evolution of PME as the head of the Gerow Board. Appointed by President Truman to evaluate the options of reorganizing the national security education community, Gerow's group generated recommendations which looked strikingly similar to the structure of today with different titles.

Goldwater, Barry Morris (1909–1998)

A stalwart conservative, the five-time Arizona Republican senator, Barry Goldwater, was a politician with a strong interest in defense reform, as a number of instances of failure proved his concerns were well founded. Born to a Jewish-Episcopalian merchant family while Arizona was still only a territory, Goldwater was an outspoken advocate for those causes he championed and he suffered fools poorly. Continuing family business while also serving as a pilot in the Air Force Reserve where he retired as a Major General, he entered local politics in the years after World War II, winning a Senate seat from a popular Democrat in 1952 with help from President Dwight Eisenhower's campaigning. Goldwater shocked many people with the bluntness and perceived radicalism of his 1964 presidential campaign. Goldwater opposed President Lyndon Johnson's civil rights legislation, overspending for social programs, foreign aid programs, and the basic approach the United States was on in 1964. He was a strong advocate of firm military action against the North Vietnamese threat as the conflict in southeast Asia expanded. President Johnson roundly defeated Goldwater but his prominent position on an array of issues was embraced by many in the United States who felt increasingly out of sync with the nation. Many analysts have subsequently noted that Goldwater's defeat in 1964 was a crucial step for Ronald Reagan's success sixteen years later, opening the door to the Republican Revolution. Goldwater won reelection to the Senate in 1968 where he served for another three terms until his 1986 retirement. His crowning piece of legislation was the Goldwater-Nichols Military Reform Act of 1986, bearing his name along with that of Alabama Congressman Bill Nichols. Goldwater was known for his personal integrity and he applied that to his beliefs of how military affairs should be run to the advantage of the nation. The military disappointments of southeast Asia between 1960 and 1975, the Iranian rescue debacle in April 1980, the Grenada miscommunications in October 1983, and even the October 1983 attack on the Marine Corps barracks in Beirut were all viewed by many in the national security establishment as indicative of a broken military system deeply needing reform. In the early 1980s, the chairman of the Joint Chiefs, Air Force General David Jones proposed reform of the military establishment that would rival the changes of the 1947 National Security Act. Goldwater remained convinced through the five years the reform package took to pass through the Congress that these reforms, including those to the PME system, were utterly necessary for the defense of the nation. Goldwater's tenacity and personal hardheadedness was a significant reason that the legislation passed, along with the hard work of a number of congressional staffers. The senator died thirteen years after his 1986 retirement back to Arizona.

Jefferson, Thomas (1743–1826)

The third president of the United States, Thomas Jefferson, had a founding role in the PME system of the United States. Jefferson was an Albermarle County, Virginia, farmer and polymath who became a Founder of the Republic during the Independence movement, from 1770 to 1783. Jefferson penned the Declaration of Independence, signed on July 4, 1776, to notify King George III of the colonists' intention to separate from British control. In the period of the negotiations over crafting the government, from 1785 to 1893, Jefferson represented the nation as envoy to France. Upon returning, Jefferson engaged in partisan politics that allowed him to defeat President John Adams as he stood for a second term in 1800. Upon taking the oath of office, Jefferson set out to greatly expand the national territory and enhance national defense. In 1803, he doubled the size of the national territory by agreeing to the Louisiana Purchase. He sent two surveyors, Merriweather Lewis and George Clark, on a historic journey to seek a northwest passage. Instead, they scouted the land to the northwest of the new nation, grasping that there was no northwest passage but a vast, unexplored and unexploited area that would require protection against foreign encroachment by European competitors. Jefferson had understood this reality when he set out the basic institution of PME with the decision to establish a military academy at West Point, New York. Overlooking the Hudson River north of the growing New York City, the academy was important to educating military officers while also producing many of the engineers and surveyors who built the United States as it expanded westward over the next century. Jefferson served two terms as president and left his mark on the nation in many ways. He died, as did his long-term rival in the political scene John Adams, on July 4, 1826, fifty years to the day after the signing of the Declaration of Independence.

Jones, David (1921–)

The chairman of the Joint Chiefs of Staff who first advocated radical military reform that became known as the Goldwater-Nichols Act, David Jones was an Air Force general who chaired the Joint Chiefs between 1978 and 1982 when he was replaced by Army General John Vessey. He was also the only chairman of the Joint Chiefs without a university degree, although he did attend the University of North Dakota and Minot State College but earned his commission during World War II when adjustments were made to normal career requirements. Jones was a controversial choice to many because he had long been labeled a "political general," a term of great derision on the part of serving officers who preferred to believe that only those choosing to act politically would do so in the United States military. Throughout his career, Jones was known for ignoring many bureaucratic norms and for focusing on his and Air Force goals instead of acquiescing to the common will. Jones had been chief of staff of the Air Force between 1974 and 1978 where he had also taken actions deemed controversial because critics charged him with being interested only in his own professional advancement, not that of the Service. A bomber pilot by rating, Jones earned the attention of the renowned (and often vilified) General Curtis LeMay in the 1950s, allowing Jones to see not only LeMay's harsh and flamboyant style but also the inner side of Robert S. McNamara's Defense Department. Jones served high in the Air Force hierarchy during the Vietnam War and the pressures to implement an all-volunteer force increased. Additionally, Jones' support for the controversial and expensive B-1 bomber led to polarized views on his skills and his common sense. Jones sought to reform the position of the chairman of the Joint Chiefs of Staff to make it a more useful advisor while also enhancing overall readiness for U.S. forces through reforming the entire structure. His ultimate success was born out with the

Goldwater-Nichols Military Reform Act of 1986, greatly strengthening PME in the United States.

Lejeune, John Archer (1867–1942)

The thirteenth commandant of the Marine Corps, John Lejeune was born in Louisiana immediately after the Civil War. He attended Louisiana State University from which he earned an undergraduate degree before attaining an appointment to the U.S. Naval Academy at Annapolis, Maryland, from which he graduated in 1888. Lejeune earned his commission and started his Marine Corps career as a second lieutenant in 1890 when he reported to Marine Barracks, New York. Between 1890 and 1907, Lejeune served primarily in the western hemisphere before moving to the Philippines for a period until 1912. Two years later Lejeune became assistant to the Marine Commandant. In 1916, Lejeune was made brigadier general, two years before becoming major general while serving in World War I in France where he won a number of distinctions. In 1919, Lejeune returned stateside to assume command over the Marine Barracks, Quantico, Virginia, followed a year later by appointment as commandant of the Marine Corps. Lejeune had a major role in reforming and expanding Marine Corps education during his period as commandant. He held the major role in standardizing education for officers at all levels, introducing schools at Quantico, and assuring that Marine Corps officers and enlisted personnel were ready for their assignments whether Corps deployments or in conjunction with other Services. He left the commandant position but stayed on active duty until the 1930s when he became superintendent of the Virginia Military Institute. Lejeune died during World War II, his mark over the Corps solid.

Luce, Stephen Bleeker (1827–1917)

Commodore Stephen Luce was the first president of the Naval War College in Newport, Rhode Island, between 1884 and 1886. Luce envisioned the institution as a focus of research and thought on the ties between technology change, naval power, and the global context in which the United States Navy operated. Luce saw the most effective manner for conducting the college's business through a balance of uniformed officers and civilian academics which would show the interaction between all aspects of the policy community. Luce's approach allowed students and faculty from the navy to better understand the tactics, operations, and strategy of a growing navy in a world where U.S. power would be more needed than ever. His most famous appointment was that of Captain Arthur Thayer Mahan, USN, to create a naval history curriculum of use to future strategists. He sought to expand thinking through the creation of the Naval Institute and the *Naval Proceedings*.[3] After his tenure as the first president, Luce returned to sea duty and returned to the college after retiring from active duty in 1890s. He died at the age of 90 in 1917.

Mahan, Arthur Thayer (1940–1914)

The son of a long-term professor of engineering at the U.S. Military Academy, Captain Alfred Thayer Mahan is easily the best-known scholar from the U.S. Navy. Rather than studying with the ground forces, this Mahan attended the U.S. Naval Academy in Annapolis, Maryland, and took a commission in the Sea Service. His volume, *The Influence of Sea Power Upon History, 1660–1805* [abridged edition (Englewood Cliffs, NJ: Prentice Hall, 1980)], remains arguably the most commonly cited book about sea power in the U.S. education system. The original 1890 publication was actually a compilation of lectures Mahan had delivered over the years. Mahan had served as a line officer on active duty in the Navy before arriving as one of the four original faculty members at Newport in 1884. Mahan's work turned the spotlight on the research and thinking of Naval War College, making it a prominent educational facility, if not the best known senior-level PME school through World War II when PME took off. Mahan's work influenced many scholars and statesmen around the world, including President Theodore Roosevelt and the German Kaiser Wilhelm II. Mahan's thesis that sea power was the basis to national power became the basis of Roosevelt's decision to launch the Great White Fleet, illustrating U.S. prowess to the world by sailing the seven seas at the turn of the twentieth century.

Mahan, Denis Hart (1802–1871)

Better know as the father of Captain Alfred Thayer Mahan, Denis Hart Mahan was an equally intellectual military officer who left an imprint on the PME system of the United States in the nineteenth century. Called upon to teach mathematics while still an undergraduate, Mahan graduated as a valedictorian of his West Point class. He took a commission in the Army but returned early in his career to the Academy to begin what evolved into a four-decade tenure at the college. His influence on the generation of officers who fought both sides of the Civil War and then executed the opening of the western portion of the nation was astounding. Mahan taught mathematics and completely revamped engineering, arguably the crucial degree program at the school. His analysis of tactics and strategy, captured in the 1847 volume *Elementary Treatise on Advance-Guard, Out-Post, and Detachment Service of Troops*, was the first U.S. book on the topic. Mahan did not receive the international acclaim of his naval-oriented son but did live to see a second son, Frederick August, graduate from the Military Academy in 1867, four years prior to the professor's death.

Marshall, George Catlett (1880–1959)

A Nobel Peace Prize winner, secretary of state, secretary of war, and Army chief of staff, General George Catlett Marshall is the man after whom the NDU headquarters building is named. Marshall's role in the twentieth century national security community is hard to capture because it was so crucial. Born in Pennsylvania on the last day of 1880, Marshall attended the Virginia Military Academy in Lexington. Marshall graduated at the top of the Corps of Cadets, earning him a commission in the U.S. Army. He went to the Philippines, where the United States was coping with its first insurgency, then received his first PME through the Army Cavalry School (1907) and Army Staff College (1908) at Fort Leavenworth, Kansas, excelling in both opportunities. Between 1908 and the War in Europe, Marshall had various staff positions. In 1917 the Army assigned him to the General Staff in France. Winning commendation for his work, Marshall was aide to Army Chief of Staff General John J. 'Blackjack' Pershing prior to being deployed in China for three years in the mid-1920s as the country descended into chaos. Upon returning, Marshall taught at the Army War College and Infantry School as well as at the Illinois National Guard. He attained the rank of brigadier general in the mid-1930s when he assumed command over the Fifth Infantry Brigade. He went to the Army Staff in 1938 and became chief of staff the following year, positioned to coordinate the massive project that became the U.S. effort in World War II. In that position, Marshall rebuilt not only the military but the educational basis for the Services, calling upon individuals such as Lieutenant Generals John L. DeWitt and Lesley J. McNair to craft the maneuvers necessary for this conflict. Marshall also managed personalities such as the fiery General Douglas MacArthur. President Franklin D. Roosevelt named Marshall, along with Dwight Eisenhower, generals of the Army, the first five-star ranks, in 1944. Marshall, in cooperation with Admiral Chester Nimitz, USN, and General Eisenhower, fought for the creation of the National War College and retooling of the Industrial College of the Armed Forces as educational institutions to cross Service and institutional barriers to better develop and implement national security strategy. After a long, distinguished career in uniform, Marshall served President Truman as special envoy to China, secretary of state, and secretary of defense prior to his retirement from public service in 1951. His name is linked with the Marshall Plan, known for its role in rebuilding Europe after its wartime devastation. In 1953, Marshall won the Nobel Peace Prize for his many efforts to promote peace around the world. He died in 1959.

MacArthur, Douglas (1880–1964)

Easily the most identifiable U.S. military officer in the Pacific theater during World War II, Douglas MacArthur also had an important role in the development of Army military education, which became a component of the PME program in place today in the United States. The son of a congressional Medal of Honor-winning, post-Civil War era army general, MacArthur spent literally his entire life in some association with the U.S. Army. His earliest years were spent in an environment where his father and fellow army soldiers defended settlers against Native American attempts to oust them from tribal lands as the United States finished its westward expansion. The younger MacArthur graduated with high distinction as first in his class of 1902 at the U.S. Military Academy in West Point, developing a strong grasp of the role that education played in the officer's view of his career. MacArthur served some of his earliest Army assignments as an aide to President Theodore Roosevelt and then General Leonard Wood. MacArthur served as a spokesperson for the Army promoting the Selective Service Act of 1917, virtually introducing the role of a public affairs office to the military and society at large with this position. MacArthur had a meteoric rise to the rank of brigadier general that he held upon embarking with the U.S. Rainbow Division in Europe during the U.S. military involvement there in 1917 and 1918. The general himself had crafted the Division from reservists and led them so successfully that he became the most highly decorated U.S. officer at the end of the War. Leaving the European theater, MacArthur assumed the position of superintendent of the Military Academy in West Point, New York, where he demanded new curriculum befitting the lessons learned from World War I. Not always patient with those who disputed him, MacArthur forced through the new thinking to create a military appropriate to the new U.S. role as a global player. Partially as a result of pressure from disgruntled Army traditionalists, Army Chief of Staff General John J. 'Black Jack' Pershing created a position in the Philippines for MacArthur to calm the environment within Army education and the Service itself. MacArthur spent a rough period of time in his personal life in the late 1920s shuttling between the U.S. colony, where he hoped to become the U.S. governor general and the continental United States where President Hoover named MacArthur Army Chief of Staff. The beginning of the 1930s proved challenging professionally as MacArthur, in his capacity as chief of staff, had to respond in 1932 to the Bonus Army troops, men who felt the nation and the Army owed them for their service, yet the Great Depression made their circumstances even worse, and MacArthur had to lead troops to roust the Bonus Army members from their protest site in the nation's capital. Three years later MacArthur accepted the invitation by Philippine nationalist leader, Manuel Quezon, to serve as head of the U.S. military delegation for the final decade of U.S. colonial rule over the Philippines, with the announced independence date of July 4, 1946 on the horizon. In 1941, MacArthur's concerns about a growing Japanese military power in East Asia were validated when the Imperial forces attacked Pearl Harbor, Singapore, and other sites along the Pacific Rim. MacArthur's Philippine forces faced defeat but the General responded to President Franklin D. Roosevelt's order to withdraw to Australia, a location from which he might be able to retake the islands. Gradually, MacArthur, in conjunction with U.S. military leaders in the continental United States, marshaled an impressive force that began retaking islands in a northward move toward Japan itself. Fulfilling his pledge of 1941, MacArthur returned to the Philippines as the final months of the war wound down. In the period after he accepted the September 1945 Japanese surrender on the *U.S.S. Missouri*, MacArthur administered the newly defeated Japan for five years, giving it a new constitution and establishing the government still in place sixty plus years later. In 1950, the aging general sought to rally United Nations forces against North Korean advances in the initial stages of the Korean War (1950–1953). While MacArthur's Inchon Landing was a bold move, his unwillingness to listen to concerns about provoking Chinese intervention led to a bloody perpetuation of the war. In April 1951, MacArthur's insubordination to President Harry S Truman led to his removal as commander in Korea, leading to a civil-military trauma in the United States when MacArthur's supporters insinuated that the General knew more than the president on this conflict. MacArthur ultimately retired from public life for the years prior to his death in 1964. His imprint on the military and overall Asian policy of the twentieth century was remarkable, capped by his innovative approach to PME at the Military Academy in West Point.

McNair, Lesley James (1883–1944)

General Lesley J. McNair, as a result of his work in the early part of World War II, is often credited with being the man who educated the Army.[4] McNair earned his Army commission upon graduating from West Point in 1904. McNair served twice in Mexico expeditions before becoming the second youngest U.S. brigadier general in World War I. After his time in Europe, McNair turned increasingly toward education, teaching at the CGSC at Fort Leavenworth, then attending the Army War College. McNair then returned to Leavenworth as the commandant where he became important to revamping the curriculum. General George Marshall put McNair in control of the Service's education and training in 1940 when Marshall became chief of staff of the General Army Headquarters. McNair's name is probably most clearly associated with the "Louisiana Maneuvers," one of three (the others being Carolina and Tennessee) sets of exercises for Army active duty and National Guard forces preparing for conflict, especially in the European theater. These exercises were comprehensive, including strategy, logistics, tactics, and all aspects of conflict, to give officers the most realistic understanding of what they would face in battle. The Louisiana Maneuvers were the beginning of Fort Polk, Louisiana, where they occurred. As a result of this training, McNair became the commander of Army Ground Forces in Europe. After the Normandy invasion in early June, he was killed by heavy bomber mistakes later that summer. Fort Lesley J. McNair, the third oldest Army post in the nation where McNair's command resided upon his death, took his name in 1948 and Congress posthumously promoted him to four-star rank to commemorate this soldier's contributions to the nation's security.

Nichols, Bill (1918–1988)

An Alabama farmer, Representative Bill Nichols shared the name of the most famous U.S. military reform act of the second half of the twentieth century with Arizona Senator Barry Goldwater. Drawn to politics and national security issues by his World War II experience, the Auburn University graduate had been part of the Army Reserve Officer Training Corps. Nichols took an Army commission as second lieutenant in 1942 and served for five years. After a decade in business, Nichols entered Alabama politics in the late 1950s, graduating to the national political scene with his election to the House of Representatives in 1966. Two years later he earned a position on the Armed Services Committee where Nichols became engaged in a number of subcommittee activities which earned him the moniker "Friend of the Serviceman." In the 1980s, Nichols was particularly bothered by the attack on the Marine Corps barracks in Beirut, which he believed could have been avoided through better coordination and Defense Department restructuring. Nichols died in office in 1988.

Nunn, Samuel Augustus (1938–)

A former senator from Georgia, Sam Nunn was one of the most committed specialists on defense issues, including education, during his twenty-four years in the nation's capital. The grandnephew of long-serving Georgia Democratic Senator Carl Vinson, Nunn served in the active duty Coast Guard in 1959–1960, followed by eight years in the Coast Guard Reserves. A lawyer, Nunn filled the seat of Richard Russell upon the latter's untimely death in late 1972. He chaired the Senate Armed Services Committee for several years, developing a well-respected expertise about military affairs. Nunn chose not to stand for election in 1996 but remained involved, with Indiana Republican Senator Richard Lugar, in trying to reduce threats of nuclear proliferation in the former Soviet Union with the "Nunn-Lugar Threat Reduction" program that bears their name. Nunn was instrumental in the crafting of the defense reforms known as Goldwater-Nichols which included substantially increased emphasis on JPME. In the late 1980s, Nunn requested a review of the curriculum at the war colleges to make certain it was increasingly rigorous to address the changing world. Nunn returned to Georgia after leaving office but retains a role in national debate on military education as a result of his ties to the Center for Strategic and International Studies in Washington, DC, a research organization which considers a range of strategic issues including PME.

Palmer, John McCauley (1870–1955)

Coming from a long line of public servants of the same name, this Army officer was crucial to the PME system in place in the United States today. Palmer earned his commission in 1892, giving him two decades' service to his nation before his major impact on the debate during World War I about national service. His posts in the Service included Ft. Leavenworth, Kansas, the China expedition, the Philippines, and the occupation of Cuba. Palmer made his mark, however, in the debate how to handle the recruitment of forces around World War I. McCauley believed forcefully in the idea of a "citizen-soldier" so he sought to see the nation push what is today known as a "Total Force" army, using the active duty military in conjunction with the Reserves and the Guard units from the states to make this a military accountable to the citizens in general. Palmer was successful as the intellectual force behind the debate in passing the National Defense Act of 1920 that created a total Army concept. From 1920 until his death thirty-five years later, Palmer pushed for better PME to enhance further the integrated military concept with the rest of society.

Richardson, James Otto (1878–1974)

This Texas Admiral was instrumental in the creation of the PME system in place in the United States today. Born in the northeast of the Lone Star State, Richardson entered Annapolis in 1898, earning his commission four years later. Lieutenant Richardson's initial assignments were in the Pacific and Atlantic fleets but he entered the initial class of the Naval Post-Graduate School in 1909. Richardson's career took him through the usual path of leadership and expertise. Richardson taught at the Naval Academy and studied at the Naval War College, which gave him a balanced understanding of military education. Richardson had several flag assignments but was relieved of command in February, prior to the Pearl Harbor attack of late 1941. Richardson remained on active duty until 1947. One of his final assignments was on the "Joint Chiefs of Staff Committee" on national military reformation. Richardson's views were well respected by virtue of his varied services. He died in 1974 at the age of 96.

Roosevelt, Theodore (1858–1919)

With President William McKinley's murder, Theodore Roosevelt assumed the presidency at the tender age of 42. While young, Roosevelt already had a varied resume and brought a virtually unending enthusiasm to the White House. Born in New York City before the Civil War, Roosevelt grew up in the "Oyster Bay" branch of the long-established Dutch family that provided two of the most important presidents in the twentieth century. His family experience proved that wealth did not inoculate people to pain: his brother, whose daughter was Eleanor Roosevelt, was a hopeless and ultimately fatal alcoholic, and both his mother and first wife died the same day in 1884, leaving him with young children. Theodore withdrew to the west to gather his thoughts and establish one of his marks on politics as an early supporter of the environment but remarried two years later, starting a second family. Long a politician, Roosevelt joined President Harrison's Civil Service Commission, then undertook to reform the New York City police department. In the 1890s, he joined the McKinley administration as assistant secretary of the Navy, implementing the ideas of Captain Arthur Thayer Mahan who believed in the role of sea power in advancing national security. With the advent of war against Spain in 1898, Roosevelt volunteered for service in Cuba and earned much national praise as the leader of the "Rough Riders." Elected governor of New York in 1898, Roosevelt proved popular and energetic but did not run for a second term when McKinley chose him as vice presidential nominee in 1900.

Upon assuming the presidency, Roosevelt proved vigorous in pursuing national security issues such as keeping U.S. interests foremost in the Caribbean Basin (known as the "Roosevelt Corollary" of the Monroe Doctrine) and pushing the U.S. Navy through deploying the Great White Fleet on a round-the-world cruise to illustrate the ability to project power. At the same time, Roosevelt sought to mediate between disputing parties in a number of differences around the world, a position for which he won the Nobel Peace Prize in 1906. Domestically, Roosevelt sought to "break" the power of the huge business "trusts," earning himself the title "Trustbuster" to the anger of many Republicans. He also worked to expand the national parks available to protect national scenery in much of the western portion of the nation. Roosevelt won reelection in 1904 but was unsuccessful in a third party run against Robert Taft in 1908. He was similarly unsuccessful as a presidential candidate in 1912. He died in his sleep in 1919. Roosevelt's contribution to PME was the appointment of Secretary of War and Secretary of State Elihu Root who radically reformed the Army. Root also increased PME by establishing the Army War College and expanded slots available at the Military Academy. Without his willingness to keep such an innovative mind, the United States would have been ill-prepared for the remainder of the twentieth century.

Root, Elihu (1845–1937)

Spanning the best part of a century, Elihu Root will always be associated with the Army War College and Roosevelt Hall at Fort Lesley J. McNair by virtue of the huge engraved plaque in the latter's rotunda bearing his name. Root followed a different path from his father, a professor of mathematics at Hamilton College in upstate New York. While he graduated first in his class, Root went on to develop a highly successful career in law, which allowed him to accumulate considerable wealth. Involved in Republican politics within New York State, President William McKinley's appointment of Elihu Root as the nominee for secretary surprised many of war. Root had no military experience but brought a freshness to the position. He implemented many new ideas that were appropriate to the new century such as the creation of an institution for Army senior officers to study the application of ground power to the issues confronting the nation in evolving its strategic situation. The Army War College was housed at the new Roosevelt Hall opened in 2003, at the confluence of the Anacostia and Potomac rivers in southwest Washington, DC. He created greater educational opportunities for special branches within the Service while also expanding the number of slots open to those seeking appointments at West Point. Root also clarified both promotion procedures and civilian control over the National Guard.[5] In sum, Secretary of War Elihu Root, along with creating an Army General Staff, proved a decisive leader for the United States at a time of transition for the nation. He returned to a private law practice briefly until President Theodore Roosevelt offered Root the secretary of state position where he excelled equally as in his prior government service. For four years as secretary of state, Root emphasized international arbitration rather than military solutions to conflict resolution. One of his most important results was the Platt Amendment that changed the Cuban Constitution to address U.S. intervention there. He ran for Senate in 1909 where he served one term. Root won the Nobel Prize for peace because of his emphasis on international negotiations. Upon retiring from the Senate, Root continued his public service by acting to represent the United States in capacities requested by President Woodrow Wilson with whom Root disagreed philosophically, but did not turn down the president's requests to represent the country where needed. Root remained active in the international relations of the nation through the mid-1920s when he was in his eighties. He served in the Carnegie Endowment for International Peace and helped establish a similar organization for Europe.

Skelton, Isaac Newton "Ike" (1931–)

A long-serving member of the House of Representatives, Missouri Democrat Ike Skelton has been personally involved in the reformulation of PME in the wake of the Goldwater-Nichols 1986 reform. Skelton, a Phi Beta Kappa graduate of the University of Missouri from the west central portion of the state, has long been interested in national security, including a personal determination in seeing academic rigor added to the curriculum. As a member of the Armed Services Committee, Skelton pushed for a smaller ratio of students to faculty and to make certain that civilian academic specialists were used instead of some sort of "old boys network" that would decrease the quality of the academic experience. Awarded honorary degrees by most PME institutions, Skelton continues pushing for higher quality professional education to promote better thinking for the nation and those who serve it under arms.

Thayer, Sylvanus (1785–1872)

The first superintendent of the U.S. Military Academy at West Point, Sylvanus Thayer was instrumental to creating the institution that has served the nation for two centuries as the oldest military educational body. Born in Braintree, Massachusetts, Thayer entered Dartmouth College but transferred to the newly opened Military Academy at West Point in 1807. Graduating a year later, Thayer's commission was in the Corps of Engineers. He served in the War of 1812. After that conflict, Major Thayer went to several European universities to understand the academic rigor applied to engineering curricula there. The Army leadership in the United States was aware of the intellectual weakness that the new Academy exhibited and Thayer was being asked to correct that. In 1817, Thayer returned to West Point as the superintendent where he dramatically overhauled the curriculum to emulate the higher standards he had seen in Europe. He created more substantial admission standards and created strong goals for graduates to attain. With his intellectual commitment to military engineering, the curriculum at West Point was heavily biased toward these subjects instead of the broader liberal arts curricula developing at small colleges around the nation. Thayer's charge, however, was to develop the nation's military officers, many of whom needed engineering to face the nation-building they were engaged in. Thayer remained as superintendent for sixteen years until he was reassigned to carry out the engineering he had long taught.[6] In the 1850s, he served briefly as the chief of the Corps of Engineers until illness prevented him from further work. Thayer died in the same town where he was born at ninety-seven years of age.

Tower, John Goodwin (1925–1991)

Once a rare species as a Republican in Texas, John Tower served as senator from 1961 until 1985. A World War II Navy veteran who continued serving in the naval reserves from 1946 until 1989,[7] Tower was a political science professor when he decided to make an unlikely run for Lyndon B. Johnson's vacant seat when the latter became John Kennedy's vice president in 1961. Prior to Tower's election, Texas appeared virtually hostile to Republican politicians back to the Civil War. Tower became a fixture in national security debates because of his knowledge of the issues and his commitment to intelligence and defense causes. He served on the Senate Armed Services Committee for several years, being instrumental in the push to add stronger standards to PME in the United States. Tower was an arms control negotiator during the Reagan administration. Further, he headed a committee, which became known by his name (the Tower Committee), to review the process that led to the Iran-Contra scandal in the second Reagan administration. The Tower Committee offered harsh criticism and several prescriptions for remedying the policy mess. President George H.W. Bush nominated Tower for secretary of defense but he never gained confirmation. He remained committed to military education questions but never served in an elected or nominated position again. He died in a plane crash in 1991.

Turner, Stansfield (1923–)

Admiral Stansfield Turner, USN, was one of the most innovative, bold thinkers in the Navy during the twentieth century. Often heavily criticized in his various positions, Turner had no doubt that there were changes needed to add rigor to the Naval War College curriculum and the work of the Central Intelligence Agency, both of which he ran. Turner graduated from the Naval Academy in Annapolis in 1947, subsequently receiving both a commission in the Navy and a Rhodes Scholarship to Oxford. Turner returned to traditional naval service but was marked as an imaginative thinker from this point. In the mid-1970s, Turner became the president of the Naval War College in Newport. He instituted a controversial curriculum of classical works, such as Herodotus, on the grounds that these works would encourage meticulous thinking, which he believed senior naval officers needed at the end of the Vietnam experience. His curriculum received considerable criticism but has led to replication elsewhere as schools in the PME system have tried to raise their standards. In 1977, former Naval officer President Jimmy Carter called upon Turner to take his new approaches to the Central Intelligence Agency where the organization was experiencing low morale as a result of investigations and upheaval of the 1970s. Turner's period at the helm was no less controversial but he shook up the traditional hierarchy of the spy community. Upon leaving office in 1981, Turner has been involved in many studies and several organizations to challenge the nation's defense and security community to think through its options and its goals. In his eighties, he remains a vibrant, challenging thinker.

Upton, Emory (1831–1881)

Emory Upton was a Civil War general who left a lasting impression on the U.S. Military Academy because of his study of tactics that remains relevant almost a hundred and fifty years after it was written. Upton originally studied at Oberlin College but transferred to West Point in the 1850s. Graduating in the top ten students of his class at the Academy, Upton received his commission the year the Civil War began.[8] He received a minor wound at the Battle of Bull Run before playing a crucial role in the Battle at Spotsylvania Court House. His sharp thought earned him the rank of brigadier general by age 24 and major general a year later. Upton was one of the shining examples of Union leadership, taking that quality with him to the position of Military Academy supserintendent after the War. He administered and taught, penning several volumes on tactics that remain standards. Upton died tragically with an incurable disease in 1881 having left a major impact on the Army educational system.

Notes

1. http://www.nps.gov/prsf/history/hrs/thompson/tt20.pdf see
2. http://www.nps.gov/prsf/history/hrs/thompson/tt20.pdf see pp. 14–16.
3. http://www.ussluce.org/lucehistory.html
4. http://www.dcmilitary.com/army/pentagram/archives/may26/pt_k52600.html
5. http://nobelprize.org/peace/laureates/1912/root-bio.html
6. http://www.bookrags.com/biography-sylvanus-thayer/
7. http://www.tsha.utexas.edu/handbook/online/articles/TT/ftoss.html
8. http://www.hollandlandoffice.com/Emory_Upton.htm

Documents

The documents selected here do not represent each and every item relating to professional military education. These are a representative sample of items with a broad range of ties to the topic.

The establishment of the Armed Forces Staff College in the Tidewater region of southeastern Virginia, already with a concentration of military units, represented a shift toward serious professional military education at the intermediate level. This is the founding order for that institution.

13 August 1946 from Secretary of the Navy—Establishment of the Armed Forces Staff College, Norfolk, Virginia

Joint Professional Military Education for the Reserve Components

One of the major lessons from the Vietnam era was the criticality of integrating the reserve component into aspects of the fight to bring the conflict to the nation in a more realistic manner. This addresses the importance of professional military education for Reservists.
From: Chairman of the Joint Chiefs of Staff
25 Aug 2004
Subject: Joint Professional Military Education (JPME) for the Reserve Components (RC)

1. Purpose. Provide information on the genesis and implementation of RC JPME and the development of Advanced Joint Professional Military Education (AJPME)

<u>E N C L O S U R E</u>

SecNav

OP24/GLB
Serial 174OP24

13 August 1946

From: Secretary of the Navy

To : All Ships and Stations

Subj: Armed Forces Staff College, Norfolk, Virginia --
       Establishment of.

1. The following college is hereby established under a
commandant and designated:

> Armed Forces Staff College
> Norfolk, Virginia   7932-400

2. The Joint Chiefs of Staff shall supervise the technical
direction of the Armed Forces Staff College.

3. The Chief of Naval Operations shall be responsible for
the operation and maintenance of the Armed Forces Staff College.

4. For purposes of administration of naval property and
naval personnel other than the faculty and students of the
Armed Forces Staff College, the following naval activity is
hereby established, under a commanding officer, and designated:

> U. S. Naval Administrative Command
> Armed Forces Staff College
> Norfolk, Virginia   7932-401

The commanding officer of this activity shall report to the
Commandant, Armed Forces Staff College, for primary duty and
shall further report for additional duty to the Commandant,
Fifth Naval District, for coordination with other naval
activities and for logistic support.

3. Bureaus and offices concerned take necessary action.

/s/ FORRESTAL

2. Key Points.
- In the Fiscal Year 1999 National Defense Authorization Act, Congress tasked the Department of Defense (DoD) to prepare RC field grade officers for joint duty assignments, by developing a course similar in content, but not identical to, the in-residence Joint Forces Staff College (JFSC) course for field grade Active Component officers. Additional congressional guidance directed that periods of in-residence training, as well as distance learning, present the best combination of academic rigor, cohort development, and cross-service acculturation.
- The Joint Forces Staff College instituted a program at the college, tasked with identifying, developing, and sustaining JPME opportunities for RC members. The resulting program is called, Reserve Component Joint Professional Military Education, or RC JPME. The first task undertaken by the RC JPME division was development of the congressionally mandated JPME course for field-grade RC officers.
- As the course is subsequent to the JPME phase-I education experienced by the students, the course was named "Advanced Joint Professional Military Education", or AJPME. This best describes its relationship within the JPME rubric as a course similar to the established JFSC JPME phase-II course.
- The AJPME curriculum is modeled on the JFSC Joint and Combined Warfighting School. Therefore, the AJPME curriculum covers National Security Systems; Command Structures; Military Capabilities; Theater (Combatant Command) Campaign Planning with Joint, Multinational, and Interagency Assets; the Joint Operation Planning and Execution System and Integration of Battlespace Support Systems.
- Technically, AJPME is a "blended" course, consisting of approximately 82 hours of advanced distributed learning (ADL) and 160 hours of face-to-face time, presented in four blocks. The curriculum is characterized by a progression from higher-order cognitive activities to affective exercises that encourages attitudinal shifts. The cognitive learning activities focus on recall and mastery of information, application of concepts and principles, problem solving, discovery, and building on existing military experiences. The affective learning activities will foster recognition and integration of new attitudes and values that promote joint acculturation.
- The inaugural AJPME class began on 29 September 2003, and graduated on 21 May 2004. Subsequent classes are on-going, and reflect the continued need for RC officers trained in joint matters. Graduates of AJPME have developed staff skills in joint and combined warfighting, emphasizing the integrated strategic deployment, operational employment, and sustainment of air, land, sea, space, and special operations forces.
- The course satisfies the requirements for advanced joint professional military education as defined in Department of Defense Instruction 1215.20, "Reserve Component (RC) Joint Officer Management Program." Officers who complete assignment to a Joint Duty Assignment - Reserve position, in accordance with the previously stated reference, are eligible for designation as "Fully Joint Qualified."
- As stated above, while the "blended" AJPME curriculum is similar to the JPME phase-II credit producing schools at JFSC, Title 10 USC, Chapter 38, Section 663, currently restricts phase-II credit to JFSC residential courses only.
- Those interested in AJPME should contact their respective RC. Each RC maintains it's [sic] own nominative process and recommends students for AJPME. For more information, visit our website at: http://www.jfsc.ndu.edu

Prepared by COL Gary P. Harper
Director, RC JPME, JFSC
INTERNATIONAL MILITARY COLLEGES
APPROVED FOR JOINT PROFESSIONAL MILITARY EDUCATION PHASE I (JPME I) EQUIVALENCE
ACADEMIC YEAR 2005–2006
Intermediate-Level JPME I Credit

- Argentine Command and Staff College (Escuela Superior de Guerre)
- Argentine Naval War College
- Argentine Air Command and Staff College
- Argentine Command and Staff College
- Australian Command and Staff College
- Belgian Command and Staff College
- Brazilian Air Force Command and Staff College
- Brazilian Army Command and Staff College (Escuela de Commando E Estado)
- Brazilian Naval War College, Command and Staff Course
- Canadian Forces Command and Staff College
- Chilean Naval War College
- Chilean Air Force Air War College (ACSC Equiv)
- Cours Superieur d'Etat Major (C.S.E.M.)
- Finnish National Defense College
- French College Interarmees de Defense (C.I.D.)
- German General Staff/Adm Staff College (Fuehrungsakademie)
- German Armed Forces Staff College
- Hellenic Air War College
- Hellenic Army War College
- India Defense Service Staff College
- Irish Command Staff College
- Italian Joint War College
- Italian Joint War College
- Instituto di Stato Maggiore Interforze (Italian War College Superior Course)
- Japanese Maritime Self Defense Forces Staff College
- Japanese Ground Self Defense Forces Staff College
- Japanese Command and Staff College
- Kuwait Joint Command and Staff College
- Republic of Korea Air Command and Staff Course
- Norwegian Armed Forces Staff College
- Norwegian Army Staff College
- Peruvian Air Command Staff College
- Royal British Air Force Staff College
- Royal (British) Joint Services Command and Staff College
- Royal Australian Joint Staff College
- Royal Australian Air Force Staff College

Even prior to the Goldwater-Nichols movement, the leadership of the Defense establishment understood the need for a program to address New flag and general officers in a joint, combined and international environment. The Capstone program accomplishes this mission.

TITLE 10 > Subtitle A > PART II > CHAPTER 38 > § 663 § 663. Education

Release date: 2005–07–12

(a) **Capstone Course for New General and Flag Officers.—**
  (1) Each officer selected for promotion to the grade of brigadier general or, in the case of the Navy, rear admiral (lower half) shall be required, after such selection, to attend a military education course designed specifically to prepare new general and flag officers to work with the other armed forces.
  (2) Subject to paragraph (3), the Secretary of Defense may waive paragraph (1)—
   (A) in the case of an officer whose immediately previous assignment was in a joint duty assignment and who is thoroughly familiar with joint matters;
   (B) when necessary for the good of the service;
   (C) in the case of an officer whose proposed selection for promotion is based primarily upon scientific and technical qualifications for which joint requirements do not exist (as determined under regulations prescribed under section 619 (e)(4) of this title); and
   (D) in the case of a medical officer, dental officer, veterinary officer, medical service officer, nurse, biomedical science officer, or chaplain.
  (3) The authority of the Secretary of Defense to grant a waiver under paragraph (2) may only be delegated to the Deputy Secretary of Defense, an Under Secretary of Defense, or an Assistant Secretary of Defense. Such a waiver may be granted only on a case-by-case basis in the case of an individual officer.
(b) **Joint Military Education Schools.—** The Secretary of Defense, with the advice and assistance of the Chairman of the Joint Chiefs of Staff, shall periodically review and revise the curriculum of each school of the National Defense University (and of any other joint professional military education school) to enhance the education and training of officers in joint matters. The Secretary shall require such schools to maintain rigorous standards for the military education of officers with the joint specialty.
(c) **Other Professional Military Education Schools.—** The Secretary of Defense shall require that each Department of Defense school concerned with professional military education periodically review and revise its curriculum for senior and intermediate grade officers in order to strengthen the focus on—
  (1) joint matters; and
  (2) preparing officers for joint duty assignments.
(d) **Post-Education Joint Duty Assignments.—**
  (1) The Secretary of Defense shall ensure that each officer with the joint specialty who graduates from a joint professional military education school shall be assigned to a joint duty assignment for that officer's next duty assignment after such graduation (unless the officer receives a waiver of that requirement by the Secretary in an individual case).
  (2)
   (A) The Secretary of Defense shall ensure that a high proportion (which shall be greater than 50 percent) of the officers graduating from a joint professional military education school who do not have the joint specialty shall receive assignments to a joint duty assignment as their next duty assignment after such graduation or, to the extent authorized in subparagraph (B), as their second duty assignment after such graduation.

(B) The Secretary may, if the Secretary determines that it is necessary to do so for the efficient management of officer personnel, establish procedures to allow up to one-half of the officers subject to the joint duty assignment requirement in subparagraph (A) to be assigned to a joint duty assignment as their second (rather than first) assignment after such graduation from a joint professional military education school.

(e) **Duration of Principal Course of Instruction at Joint Forces Staff College.—**
   (1) The duration of the principal course of instruction offered at the Joint Forces Staff College may not be less than three months.
   (2) In this subsection, the term "principal course of instruction" means any course of instruction offered at the Joint Forces Staff College as Phase II joint professional military education.

Only a portion of the next document appears because the guidance on enlisted professional military education is less developed than that for officers. This is the equivalent of the OPMEP, the *Officer Professional Military Education Program*.

Enlisted Professional Military Education Policy

3. *Intent*
   a. Professional development is the product of a learning continuum that comprises individual training, experience, education, and self-improvement. The role of PME is to provide the education needed to complement individual training, operational experience, and self-improvement to produce the most technically proficient, professionally competent, and self-confident individual possible. Within our enlisted ranks, the focus of learning opportunities centers on individual training (how to do). As enlisted personnel grow in experience and assume greater responsibilities, individual training is enhanced with professional education (how to think) opportunities.

Service PME

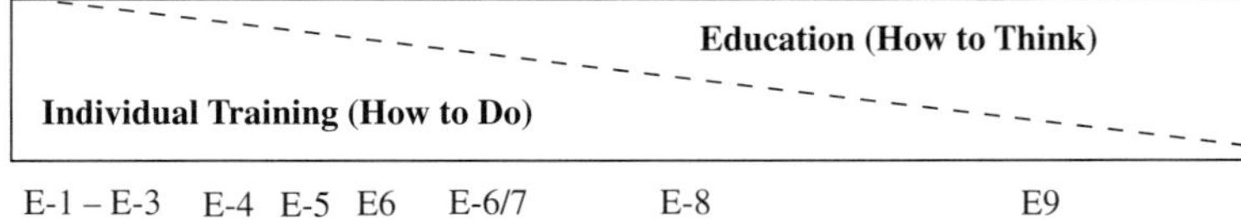

   b. In its broadest conception, education conveys general bodies of knowledge and develops habits of mind applicable to a broad spectrum of endeavors. At its highest levels and in its purest form, education fosters breadth of view, diverse perspectives, critical analysis, abstract reasoning, comfort with ambiguity and uncertainity, and innovative thinking, particularly with respect to complex, non-linear problems. This contrasts with training, which focuses on the instruction of personnel to enhance their capacity to perform specific functions and tasks.

c. Training and education are not mutually exclusive. Virtually all military schools and professional development programs include elements of both education and training in their academic programs. Achieving success across the learning continuum relies on close coordination and integration of training and education to develop synergies as personnel develop over time, acquiring and performing progressively more complex and demanding skills and responsibilities as they advance in their careers.

5. *Definitions.* See Glossary.

6. *Responsibilities*
a. Per reference a, the Chairman is responsible for formulating policies for coordinating the military education and training of members of the armed forces.
b. Enclosure A outlines the policies and procedures necessary to fulfill CJCS PME vision and responsibilities for the enlisted force. Enclosures B through D address specific EPME policies and provide guidance to Service Chiefs on joint emphasis areas, which consist of joint learning areas (JLAs) and joint learning objectives (JLOs) that should be included in Service EPME programs. Enclosure E outlines JLAs and JLOs that define the EJPME programs. Enclosure F is a list or references pertaining to this instruction.

7. *Summary of Changes.* N/A

8. *Releasability.* This instruction is approved for public release: distribution is unlimited. DOD components (to include the combatant commands), other federal agencies, and the public amay obtain copies of this instruction through the Internet from the CJCS Directives Home page—http://www.dtic.mil/cjcs_directives. Copies are also available through the Government Printing Office on the Joint Electronic Library CD-ROM.

9. *Effective Date.* This instruction is effective for planning and programming upon receipt.

PETER PACE
General, United States Marine Corps
Chairman
of the Joint Chiefs of Staff

A key aspect of any task for the military is understanding the commander's intent. In this instance, grasping the secretary of defense's and chairman of the Joint Chiefs of Staffs statutory and regulatory responsibilities is crucial. This appears from the Joint Staff J-7 (Education and Training) Web site at http://www.au.af.mil/au/awc/awcgate/mil-ed/cjcs-pme-responsibilities.pdf (accessed on March 13, 2006}

Military Education CJCS Responsibilities by Law

- Formulates policies for coordinating the military education and training of members of the armed forces *(10 USC, Section 153 (a) (5) (C))*
- Advises and assists the Secretary of Defense by periodically reviewing and revising the curriculum of each school of the National Defense University and of any other joint professional military education school to enhance the education and training of officers in joint matters. *(10 USC, Section 663 (b))*
- Advises and assists the Secretary of Defense with promulgation of cost-accounting system for use by Military Departments in preparing budget requests for operation of professional military schools *(10 USC, Section 2162 (a))**
- Advises Military Departments on NDU's budget needs *(10 USC, Section 2162 (b)(2))*
- "The Secretary of Defense shall require that each Department of Defense school concerned with professional military education periodically review and revise its curriculum for senior and intermediate grade officers in order to strengthen the focus on joint matters and preparing officers for joint duty assignments" *(10 USC, Sec 663 (c))**

Military Education Chairman of the Joint Chiefs of Staff Responsibilities by Regulation

- Approves charter and mission of the National Defense University (NDU) and its components
  - Recommends National Defense University President nominee to Secretary of Defense
  - Selections National War College, Industrial College of the Armed Forces, and Joint Forces Staff College Commandants
- Accredits the Program of Joint Education at Service and NDU colleges
- Approves nominees for CJCS Chairs at Service and NDU colleges
- Approves the annual list of JPME Phase I-equivalent international military colleges
- Selects countries to send officers to the International Fellows Program at NDU

This is an extensive document laying out the goals, responsibilities, and other aspects of the Officer Professional Military Education Policy (OPMEP) that is regularly revised by the J-7 in consultation with many other parts of the U.S. government to assure its timeliness.

These are President Ronald Reagan's remarks on signing the Goldwater-Nichols bill in 1986.

Message to the Congress Outlining Proposals for Improving the Organization of the Defense Establishment

April 24, 1986
To the Congress of the United States:
On February 26, I spoke to the American people of my highest duty as President—to preserve peace and defend the United States. I outlined the objectives on which our defense program has rested. We have been firmly committed to rebuilding America's strength, to meeting new challenges to our security, and

---

*Secretary of Defense responsibilities delegated to Assistant Secretary of Defense for Financial Management and Planning in Secretary of Defense memo of 2 June 1990.

to reducing the danger of nuclear war. We have also been dedicated to pursuing and implementing defense reforms wherever necessary for greater efficiency or military effectiveness.

With these objectives in mind, I address the Congress on a subject of central importance to all Americans—the future structure and organization of our defense establishment.

Extensive study by the Armed Services Committees of the Senate and the House of Representatives has produced numerous proposals for far-reaching changes in the structure of the Department of Defense, including the organization of our senior military leadership. These proposals, sponsored by members with wide knowledge and experience in defense matters, are now pending before the Congress.

In addition, a few weeks ago I endorsed the recommendations of the bipartisan President's Blue Ribbon Commission on Defense Management, chaired by David Packard, for improving overall defense management including the crucial areas of national security planning, organization, and command.

For more effective direction of our national security establishment and better coordination of our armed forces, I consider some of these proposals to be highly desirable, and I have recently taken the administrative steps necessary to implement these improvements. In this message, I wish to focus on the essential legislative steps that the Congress must take for these improvements to be fully implemented.

Together, the work of the Packard Commission and the Congress represents certainly the most comprehensive review of the Department of Defense in over a generation. Their work has been the focus of an historic effort to help chart the course we should follow now and into a new century. While we will continue to refine and improve our defense establishment in the future, it will be many years before changes of this scope are again considered. Given these unique circumstances, I concluded that my views as President and Commander in Chief should be laid before the Congress prior to the completion of legislative action.

Executive and Legislative Responsibilities

In forwarding this message, I am cognizant of the important role of the Congress in providing for our national defense. We must work together in this endeavor. However, any changes in statute must not infringe on the constitutionally protected responsibilities of the President as Commander in Chief. Any legislation in which the issues of Legislative and Executive responsibilities are confused would be constitutionally suspect and would not meet with my approval.

My views concerning legislation on defense reorganization now pending in the House and Senate reflect a reasoned and open-minded approach to the issues, while maintaining a close watch on the constitutional responsibilities and prerogatives of the Presidency. While I had considered forwarding a separate bill to the Congress, I concluded that this was not necessary since many of the legislative recommendations of the Packard Commission are already pending in one or more bills. However, additional changes in law are also proposed in those other bills, and such changes must be carefully weighed.

Certain changes in the law are necessary to accomplish the objectives we seek. Among these are the designation of the Chairman of the Joint Chiefs of Staff as the principal military adviser to the President, the Secretary of Defense, and the National Security Council, and the Chairman's exclusive control over the Joint Staff; the creation of a new Vice Chairman of the Joint Chiefs of Staff; and the creation of a new Level II position of Under Secretary of Defense for Acquisition.

Other proposed changes in law are, in my judgment, not required. It is not necessary to place in law those aspects of defense organization that can be accomplished through executive action. Nevertheless, if such changes are recommended by the Congress, I will carefully consider them, provided they are consistent with current policy and practice and do not infringe upon the authority or reduce the flexibility of the President or the Secretary of Defense.

General Principles

The organization of our present-day defense establishment reflects a series of important reforms following World War II. These reforms were based upon the harsh lessons of global war and were hastened by the new military responsibilities and threats facing our Nation. They culminated in 1958 with the reorganization of the Department of Defense under President Eisenhower.

President Eisenhower's experience of high military command has few parallels among Presidents since George Washington. The basic structure for defense that he laid down in 1958 has served the Nation well for over 25 years. The principles that governed his reorganization proposals are few but fundamental. They are of undiminished importance today.

First, the proper functioning of our defense establishment depends upon civilian authority that is unimpaired and capable of strong executive action.

As civilian head of the Department, the Secretary of Defense must have the necessary latitude to shape operational commands, to establish clear command channels, to organize his Office and Department of Defense agencies, and to oversee the administrative, training, logistics, and other functions of the military departments.

Second, if our defense program is to achieve maximum effectiveness, it must be genuinely unified.

A basic theme of defense reorganization efforts since World War II has been to preserve the valuable aspects of our traditional service framework while nonetheless achieving the united effort that is indispensable for our national security. President Eisenhower counseled that separate "service responsibilities and activities must always be only the branches, not the central trunk of the national security tree."

Unified effort is not only a prerequisite for successful command of military operations during wartime, today, it is also indispensable for strategic planning and for the effective direction of our defense program in peacetime. The organization of our senior military leadership must facilitate this unified effort. The highest quality military advice must be available to the President and the Secretary of Defense on a continuing basis. This must include a clear, single, integrated military point of view. Yet, at the same time, it must not exclude well-reasoned alternatives.

Third, the character of our defenses must keep pace with rapid changes in the military challenges we face.

President Eisenhower observed a revolution taking place in the techniques of warfare. Advancing technology, and the need to maintain a vital deterrent, continually test our ability to introduce new weapons into our armed forces efficiently and economically. It is increasingly critical that our forces be able to respond in a timely way to a wide variety of potential situations. These range across a spectrum from full mobilization and deployment in case of general war, to the discriminating use of force in special operations. To respond successfully to these changing circumstances and requirements, our defense organization must be highly adaptable.

Where the roles and responsibilities of each component of our defense establishment are necessarily placed in law, they must be clear and unambiguous, but not so constrained or detailed as to impair operational flexibility or the common sense of those in positions of responsibility. Laws must not be written in response to the strengths and weaknesses of individuals who now serve. Instead, they should establish sound, fundamental relationships among and between civilian and military authorities, relationships that reflect the proper balance between our traditions and heritage and the practical considerations unique to military matters.

Special Relationships Between the President and Certain Subordinates

I noted earlier that President Eisenhower brought to his Presidency a unique perspective and unprecedented military experience. Few Presidents have come into this office as well prepared as he to assume the responsibilities of Commander in Chief. This fact places a heavy burden on our defense establishment and requires the continued development of key institutions and relationships that constitute the framework of our current organization.

It has been my experience that within this framework there is a special relationship between the President, the Secretary of Defense, and the Combatant Commanders. In providing for the timely and effective use of the armed forces in support of our foreign policy, our entire defense establishment is focused on supporting this special relationship and making it as effective as possible. All other aspects of our defense organization must be subordinate to this purpose.

The Secretary of Defense. In particular, the law places broad authority and heavy responsibilities on the Secretary of Defense. The Secretary, in his responsibility as head of the defense establishment and in executing the directives of the Commander in Chief, embodies the concept of civilian control. No one but the President of the United States and the Secretary of Defense is empowered with command authority over the armed forces. In managing the Department of Defense the Secretary must retain the authority and flexibility necessary to fulfill these broad responsibilities.

Thus, where the Congress seeks statutory changes that would affect the Secretary of Defense, I will apply the following criteria:

- I will support efforts to strengthen the authority of the Secretary of Defense if there are areas in the law where his current authority is not sufficiently clear.
- The Secretary's authority should be delegated as he sees fit, and such delegation should never be mandated in the law apart from his concurrence and approval.
- The strengthening of other offices or components of the defense establishment should never be, nor appear to be, at the expense of the authority of the Secretary of Defense.

The Combatant Commanders. The Unified and Specified Commanders are the individuals in whom the American people and our defense establishment place warfighting responsibilities. The Secretary and I consult the Combatant Commanders for their joint and operational points of view in determining how our military forces should be used and in determining our military requirements for important geographic and functional areas. Their successes in any future conflict would depend in large measure on how well we plan for their needs in today's defense budgets.

With this in mind, the Secretary initiated regular meetings with the Combatant Commanders and has provided them greater access to the Department's internal budget process. In addition, I am implementing the recommendations of the Packard Commission to improve the channel of communications between the President, the Secretary, and the Combatant Commanders; to provide broader authority to those Commanders to structure their subordinate commands; to provide options in the organizational structure of Combatant Commands for the shortest possible chains of command consistent with proper supervision and support; and to provide for flexibility where issues or situations overlap the current geographical boundaries of the Combatant Commands.

These changes reflect an evolutionary and positive trend toward strengthening the role of the operational commanders within the defense establishment. While I hope and expect this trend will continue, it is not necessary that these efforts be mandated in the law. If the Congress wishes to elaborate on the current law, there are several important issues that should be considered:

- In organizing our forces to maximize their combat potential under a variety of circumstances, the President and Secretary of Defense must retain the authority for establishing Combatant Commands; for prescribing their force structure; and for oversight of the assignment of forces by the Military Departments. To be effective, this authority requires broad latitude and flexibility and calls for a minimum amount of statutory constraint. Restrictions in the law that prohibit the establishment of certain command arrangements should be repealed. My authority as Commander in Chief is sufficient to deal with any necessary command arrangements or adjustments in the assignment of forces that unforeseen circumstances could require.

- In moving to strengthen the role of the Combatant Commanders we must establish an appropriate balance between enhancing their influence in resource allocation and maintaining their focus on joint training and operational planning. The Combatant Commanders must have sufficient authority and influence to accomplish their mission, within the constraints necessarily established by the Secretary, without being burdened with administrative responsibilities that detract from their primary role as operational commanders.
- Finally, we must not legislate departmental procedures. The changes I have initiated concerning the defense planning and budgeting process provide for the further development of the role of the Combatant Commanders. It is neither necessary nor appropriate for the Department's internal resource allocation process to be defined in law. The establishment and evolution of such procedures must remain the prerogative of the Secretary of Defense.

The Chairman of the Joint Chiefs of Staff. In the relationship between the President, the Secretary of Defense, and the Combatant Commanders, there is a special role for the Chairman of the Joint Chiefs of Staff. The Chairman ranks above all other officers and devotes all of his time to joint issues. I deal with him or his representative on a regular basis and he serves as the primary contact for the Secretary and me on operational military matters. As a matter of practice, the Chairman also functions within the chain of command by transmitting to the Combatant Commanders those orders I give to the Secretary. Under the directive I recently signed to implement the recommendations of the Packard Commission, this practice will be broadened and strengthened.

In this regard, I have concluded that the Chairman's unique position and responsibilities are important enough to be set apart and established in law, and that he should be supported by a military staff responsive to his own needs and those of the President and the Secretary of Defense. In reaching this judgment I have carefully weighed the view that concentration of additional responsibility in the Chairman could limit the range of advice provided to me and the Secretary, or somehow undermine the concept of civilian control. While this concern is understandable, it does not apply to the structural changes I would endorse. Since the Chairman and the Joint Chiefs of Staff will continue to function together as military advisors and the Secretary's military staff, and the Chairman will continue to report directly to the President and the Secretary of Defense, none of the new responsibilities of the Chairman that I propose would diminish the authority or control of the Secretary of Defense. Accordingly, I support legislation that will accomplish the following objectives:

- Designate the Chairman of the Joint Chiefs of Staff as the principal uniformed military advisor to the President, the National Security Council, and the Secretary of Defense;
- Place the Organization of the Joint Chiefs of Staff and the Joint Staff under the exclusive direction of the Chairman, to perform such duties as he prescribes to support the Joint Chiefs of Staff and respond to the President and the Secretary of Defense; and
- Create the new position of Vice Chairman of the Joint Chiefs of Staff and make the Vice Chairman a member of the Joint Chiefs of Staff.

While recognizing and providing for the special role of the Chairman in the law, the basic structure of the Joint Chiefs of Staff should be retained. The advantages and disadvantages of the current system, in which the Chiefs of the Services provide advice concerning both their military Service and joint issues, have been debated for many years and are well known. I believe that certain disadvantages will be remedied by a stronger Chairman without sacrificing the advantages of the current system. I find that the Chiefs of the Services are highly knowledgeable regarding particular military capabilities. And, just as important, joint military perspectives on both resource allocation and operations, developed under the Chairman's leadership, must be upheld and supported at the highest levels of the Military Departments.

For these reasons, as we take the appropriate steps to strengthen the role of the Chairman, the law must ensure that:

- The Service Chiefs remain members of the Joint Chiefs of Staff; and that, in addition to the views of the Chairman, the President is also provided with the views of other members of the Joint Chiefs of Staff.
- In addition, in creating the new position of Vice Chairman, the law must provide flexibility for the President and Secretary of Defense to determine who shall serve as Acting Chairman in the Chairman's absence.

In our efforts to strengthen the ability of the Chairman and the Joint Chiefs of Staff to be responsive to the civilian leadership, we must also make certain that the military establishment does not become embroiled in political matters. The role of the Chairman and other members of the Joint Chiefs of Staff is strictly advisory in nature and, with the armed forces as a whole, they serve the American people with great fidelity and dedication. In my view, changes in the tenure of the Chairman or other senior officers that are tied to the civilian electoral process would endanger this heritage. I oppose any bill whose provisions would have the effect of politicizing the military establishment.

Acquisition Reform

The Packard Commission has pointed out what we all know to be true: that our historic ups and downs in defense spending have cost us dearly over the long term. For many years there has been chronic instability in both top-line funding and individual programs. This has eliminated key economies of scale, stretched out programs, and discouraged defense contractors from making long-term investments required to improve productivity. To end this costly cycle we must find ways to provide the stability that will allow the genius of American ingenuity and productivity to flourish.

We also know that Federal law governing procurement has become over-whelmingly complex. Each new statute adopted by the Congress has spawned more administrative regulation. As laws and regulations have proliferated, defense acquisition has become ever more bureaucratic and encumbered by over-staffed and unproductive layers of management. We must both add and subtract from the body of law that governs Federal procurement, cutting through red tape and replacing it with sound business practices, innovation, and plain common sense.

The procurement reforms I have begun within the Executive branch cannot reach their full potential without the support of the Congress. We must work together in this critical period, where so many agree that our approach to defense procurement in both the Executive and Legislative branches is in need of repair. However, in moving forward to implement needed reforms, I urge the Congress to show restraint in the use of more legislation as a solution to our current problems.

The Commission identified the need for a full-time defense acquisition executive with a solid industrial background. This executive would set overall policy for procurement and research and development, supervise the performance of the entire acquisition system, and establish policy for the oversight of defense contractors. I concur with this recommendation.

— The Congress should create by statute the new Level II position of Under Secretary of Defense for Acquisition through the authorization of an additional Level II appointment in the Office of the Secretary of Defense.
— Beyond this initiative, however, further change to the acquisition organization of the Department of Defense should be left to the Executive branch. The procurement reforms I have recently set in motion are fundamental and far-reaching and should be allowed to proceed without the burden of further piece-meal changes. I call on the Congress to demonstrate restraint in two particular areas:
— First, with the exception of changes to procurement or anti-fraud laws I have already endorsed, we should refrain from further action to add new procurement laws to our statutes pending the complete review of all Federal statutes governing procurement that I have recently directed. The vast body of procurement

law that now exists must be simplified, consolidated, and made responsive to our national security needs.
— And second, we should take no further action to add new laws that would restrict the authority of the Secretary of Defense to hire and retain the high quality of personnel needed to administer the Department of Defense's acquisition program.

If citizens from the private sector who participate in the conduct of government are unfairly prohibited from returning to their livelihood, it will not be just their willingness to serve that will suffer. The Nation will suffer as well. I will later report to the Congress on steps I am taking or that I propose the Congress take in these areas. And I will also review and report on the accountability of the defense industry to the Department of Defense, and to the American people. This review will address the ethics of the industry, the Department of Defense's oversight responsibility, and the role of the Department's Inspector General. I urge the Congress not to act in these important areas until it has had an opportunity to review my report.

While the Department of Defense and Executive branch are focused on implementing the details of these reforms, I urge the Congress to focus its attention on the structural and procedural reforms that are also essential for the stability we seek.

Two-year defense budgets are an essential step toward stability. I urge the Congress to develop internal procedures for the authorization and appropriation of defense budgets on a biennial basis, beginning with the FY 1988 budget. My FY 1988 defense budget will be structured with this in mind.

The Congress should encourage the use of multiyear procurement where appropriate on a significantly broader scale. Multiyear procurement is a strong force for stability and efficiency. We have already saved billions of dollars through multiyear procurement and have never broken a contract or suffered a single loss to date. We want to continue and expand our efforts in this important area.

Milestone funding of research and development programs is also a form of multiyear contracting. I will work with the Congress to select appropriate programs to be base-lined in cost over a multiyear period so that these programs can be funded in an orderly and stable fashion. If we know what we want to accomplish, we can set a proper ceiling on costs and manage our program within those costs. I urge the Congress to support milestone funding and the base-lining concept of placing a ceiling on research and development costs.

Finally, there are some forty different committees or subcommittees that claim jurisdiction over some aspect of the defense program. This fragmented oversight process is a source of confusion, and it impedes the cooperation between the Congress and the Executive branch so necessary to effective defense management. I urge the Congress to return to a more orderly process involving only a few key committees to oversee the defense program. Only with such reform can we achieve the full benefits of those changes now underway within the Department of Defense.

Working together, we have accomplished a great deal over the past five years. Yet there is more to be done. This effort represents a new beginning for our defense establishment. When these reforms have been achieved we will have:

— developed a rational process for the Congress and the President to reach enduring agreement on national military strategy, the forces to carry it out, and the stable levels of funding that should be provided for defense;
— strengthened the ability of the military establishment to provide timely and integrated military advice to civilian leadership;
— improved the efficiency of the defense procurement system and made it more responsive to future threats and technological needs; and
— reestablished the bipartisan consensus for a strong national defense.

The Packard Commission has charted a three-part course for improving our Nation's defense establishment. I have already directed implementation of its recommendations where that can be accomplished through Executive action. In this message, I ask that the Congress enact certain changes in law that will further improve the organization and operation of the Department of Defense. Now, the remaining requirement for reform lies within the Congress itself.

I began this message by emphasizing the important role of Congress in our defense establishment. In the organizational changes we now address, the Congress should be commended for fulfilling its broad responsibility to make laws to organize and govern the armed forces. However, with respect to the changes we must consider in the areas of budget, resource allocation, and procurement, the future is much less certain. To establish the stability essential for the successful and efficient management of our defense program, the Congress must be more firmly committed to its constitutional obligations to raise and support the armed forces.

Within the limits of my authority as President, I will continue to improve and refine the national security apparatus within the Executive branch. And I will support any further changes in procedures, regulations, or statutes that would improve the long-term stability, effectiveness, and efficiency of our defense effort.

In having fully committed ourselves to implementing the Packard Commission's recommendations, this Administration has overcome the difficult bureaucratic terrain that has stood in the path of previous efforts. Now, we face a broad ocean of necessary congressional reforms in which the currents of politics and jurisdiction are equally treacherous. We must not stop at the water's edge.

Only meaningful congressional reform can complete our efforts to strengthen the defense establishment and develop a rational and stable budget process—a process that provides effectively and efficiently for America's security over the long haul.

With a spirit of cooperation and bipartisanship, confident that we can rise to this occasion, I stand ready to work with the Congress and meet the challenge ahead.

Ronald Reagan

The White House,

April 24, 1986.

CHAIRMAN OF THE JOINT CHIEFS OF STAFF INSTRUCTION

J-7 CJCSI 1800.01C
OFFICER PROFESSIONAL MILITARY EDUCATION POLICY (OPMEP).

1. Purpose. This instruction distributes the policies, procedures, objectives, and responsibilities for officer professional military education (PME) and joint officer professional military education (JPME). Chairman of the Joint Chiefs of Staff (CJCS) authority derived from title 10, USC, section 153(a)(5)(C).

2. Cancellation. CJCSI 1800.01B, 30 August 2004, "Officer Professional Military Education Policy," is canceled.

3. Applicability. This instruction applies to the Joint Staff, the National Defense University (NDU), and the Military Services. It is distributed to other agencies for information only.

4. Chairman's Vision

   a. PME–both Service and Joint–is the critical element in officer development and is the foundation of a joint learning continuum that ensures our Armed Forces are intrinsically learning organizations. The PME vision understands that young officers join their particular Service, receive training, and education in a joint context, gain experience, pursue self development, and over the breadth of their careers, become the senior leaders of the joint force. Performance and potential are the alchemy of this growth, but nothing ensures that they are properly prepared leaders more than the care given to the content of their training, education, experience, and self-development opportunities. My PME vision entails ensuring that officers are properly prepared for their leadership roles at every level of activity and employment, and through this, ensure that the US Armed forces remain capable of defeating today's threat and tomorrow's.

   b. Today, the United States enjoys an overwhelming qualitative advantage not only in our fielded capabilities, but in our cognitive approach to our duties; sustaining and increasing this advantage will require a transformation achieved by combining technology, intellect, and cultural changes across the joint community. PME needs to continue to build an officer that understands the strategic implications of tactical actions and the consequences that strategic actions have on the tactical environment. Service delivery of PME, taught in a joint context, instills basic Service core competencies; JPME instills joint core competencies. JPME should position an officer to recognize and operate in tactical, operational, and strategic levels of national security.

   c. The legislative changes dictated in the Ronald W. Reagan National Defense Authorization Act of 2005 have expanded the opportunities to receive JPME and established a link between joint officer development and JPME). The future joint force requires knowledgeable, empowered, innovative, and decisive leaders capable of succeeding in fluid and perhaps chaotic operating environments with more comprehensive knowledge of interagency and multinational cultures and capabilities. This policy document is at the heart of building those officers.

d. As always, the men and women of our Armed Forces are the nation's most important strategic resource. Only a force of dedicated, highly educated, and well-trained men and women capable of leveraging new ideas will succeed in the complex and fast-paced environment of future military operations. Moreover, this force must exhibit honor integrity, competence, physical and moral courage, dedication to ideals, respect for human dignity, the highest standards of personal and institutional conduct, teamwork, and selfless service. Thus, it is imperative to maintain sustained emphasis on ethical conduct and the highest ideals of duty, honor, and integrity at all PME and JPME institutions.

5. Responsibilities

a. The Chairman, as defined by law, is responsible for the following tasks related to military education: (1) Formulating policies for coordinating the military education and training of members of the Armed Forces (subparagraph (a)(5)(C), reference a); (2) Advising and assisting the Secretary of Defense (SecDef) by periodically reviewing and revising the curriculum of each school of NDU (and of any other JPME school) to enhance the education and training of officers in joint matters (section 2152, paragraph (b), reference b); and (3) Advising and assisting the SecDef through the designation and certification of all elements of a JPME (Phase I, II and CAPSTONE (section 2154, paragraph (a), reference (b) . . .

b. Adds Chairman's responsibility to advise and assist the SecDef through the designation and certification of all elements of a joint professional military education—consistent with the 2005 National Defense Authorization Act (NDAA).

c. Delineates which schools and colleges have authority to teach JPME Phase II and allows for completion of JSO educational requirements by graduating from a JPME Phase II accredited Senior Level College (SLC).

d. Modifies Joint Learning Areas (JLA) and Joint Learning Objectives (JLO) to be consistent with 2005 NDAA language as it relates to mandated subject areas for JPME Phase I and II.

e. Standardizes terminology for "single-phase JPME" vice use of the term "full-JPME." Provides a definition for single-phase JPME in the glossary.

f. Adds accreditation dates and levels for AJPME and JAWS.

g. Mandates class mix at each Service Senior Level College (SLC) shall have no more than 60 percent host Military Department students.

h. Clarifies computations of Class and Seminar mix through inclusion of US military officers, international officers and civilian enrollments in the student body.

i. Mandates total non-host Service SLC military permanent faculty shall be no less than 40 percent of the total military faculty.

j. Defines non-host Military Department faculty as those whose primary duty is student instruction of JPME.

k. Changes Service SLC faculty mix requirements from a mandated 10 percent to a proportional division among each non-host Military Department.

i. Clarifies approaches used to provide non-resident JPME and provides appropriate definitions in the glossary.

j. Adds the Distance Learning Coordination Committee and General and Flag Officer Coordination Committee as subgroups of the Military Education Coordination Council Working Group.

k. Modifies Service Chief responsibilities relative to non-host Service SLC student and faculty mixes.

l. Replaces the descriptive verbs in the illustrative level of Appendix A to Enclosure E, Learning Objective Verbs, for "Value".

m. Includes JLOs that address combating weapons of mass destruction/effects (WMD/E). Provides a definition in the glossary.

n. Incorporates a JLO that addresses cultural awareness for Primary, ILC, SLC, NWC, ICAF, JCWS, JAWS, and AJPME.

o. Adds an appendix for Phase II JPME at in-residence Service SLCs. Annotates appropriate JLAs and JLOs.

p. Adds a JLA and supporting JLOs for information operations under the JAWS.

q. Adds an appendix for Functional Component Commanders' Courses and establishes appropriate JLAs and JLOs.

r. Modifies JLAs and JLOs under JFOWC consistent with CAPSTONE, PINNACLE and the Functional Component Commanders' Courses.

s. Updates the 6-year Process for Accreditation of Joint Education (PAJE) schedule.

t. Updates the references, glossary and definitions.

u. Establishes CAPSTONE as part of the three-phased approach to JPME.

v. Establishes Supreme Allied Commander Transformation (SACT) as authority to issue invitations for allied participation in the PINNACLE JOM.

7. Releasability. This instruction is approved for public release; distribution is unlimited. DOD components (to include the combatant commands), other federal agencies, and the public may obtain copies of this instruction through the Internet from the CJCS Directives Home Page— http://www.dtic.mil/cjcs_directives. Copies are also available through the Government Printing Office on the Joint Electronic Library CD-ROM.

8. Effective Date. This instruction is effective for planning and programming upon receipt. Colleges and schools have 1 year to meet new guidelines.

9. Revisions. Submit recommended changes to this policy to the Joint Staff, J-7, Joint Education Branch, 7000 Joint Staff, Pentagon, Washington, D.C. 20318-7000.

10. Information Requirements. Reports required by this policy are exempt from normal reporting procedures in accordance with referencee.

For the Chairman of the Joint Chiefs of Staff:
Approved & Secured with ApproveIT
by: WALTER L. SHARP, 22 December 2005
WALTER L. SHARP
Lieutenant General, USA
Director, Joint Staff

TABLE OF CONTENTS

ENCLOSURE A

OFFICER PROFESSIONAL MILITARY EDUCATION POLICY

1. Overview. The Officer Professional Military Education Policy (OPMEP) defines CJCS objectives and policies regarding the educational institutions that comprise the officer PME and JPME systems. The OPMEP also identifies the fundamental responsibilities of the major military educational participants in achieving those objectives.

   a. The Services and NDU provide officer PME and JPME to members of the US Armed Forces, international officers, eligible federal government civilians, and other approved students.

      (1) Services operate officer PME systems to develop officers with expertise and knowledge appropriate to their grade, branch, and occupational specialty. Incorporated throughout Service-specific PME, officers receive JPME from pre-commissioning through G/FO level.

      (2) NDU institutions enhance the education of selected officers and civilians in national security strategy, resource management, information resources management, information operations and joint and multinational campaign planning, and warfighting.

   b. All officers should make a continuing, strong personal commitment to their professional development beyond the formal schooling offered in the military educational system. Officers share responsibility for ensuring continued growth of themselves and others.

2. Scope. This instruction addresses PME and JPME from precommissioning to G/FO levels.

3. Intent

   a. Professional development is the product of a learning continuum that comprises training, experience, education, and self-improvement. PME provides the education needed to complement training, experience, and self-improvement to produce the most professionally competent individual possible.

   b. In its broadest conception, education conveys general bodies of knowledge and develops habits of mind applicable to a broad spectrum of endeavors. At its highest levels and in its purest form, education fosters breadth of view, diverse perspectives and critical analysis, abstract Enclosure A reasoning, comfort with ambiguity and uncertainty, and innovative thinking, particularly with respect to complex, non-linear problems. This contrasts with training, which focuses on the instruction of personnel to enhance their capacity to perform specific functions and tasks.

   c. Training and education are not mutually exclusive. Virtually all military schools and professional development programs include elements of both education and training in their academic programs. Achieving success across the joint learning continuum relies on close coordination of training and education to develop synergies as personnel develop individually over time, acquiring, and performing progressively higher skills and responsibilities as their careers advance.

   d. Opportunities for substantial professional education are relatively rare—particularly for the extended in-residence education that produces a synergy of learning that only come from daily, face-to-face interaction with fellow students and faculty. Consequently, the PME institutions should strive to provide as pure and high quality education as feasible.

4. Training Transformation (T2)

   a. On 1 March 2002, the Department of Defense issued a *Strategic Plan for Transforming DOD Training* to provide dynamic, capabilities-based training in support of national security across the full spectrum of Service, joint, interagency, intergovernmental, and multinational operations. Key objectives of T2 include preparing individuals to: think intuitively joint; improvise and adapt to emerging crises; and achieve unity of effort from diversity of means to meet the joint operational requirements of the combatant commanders. T2 regards joint education as fundamental to creating a culture that supports transformation, founded on leaders who are innately joint, and comfortable with change. T2 requires joint education to prepare leaders both to conduct operations as a coherently joint force and to think their way through uncertainty.

b. T2 efforts have implications for military education. The CJCS, as advised by the Director, Joint Staff, the Deputy Director of the Joint Staff for Military Education (DDJS-ME), and the Military Education Coordination Council (MECC, see Enclosure D), retains responsibility for formulating policies for coordinating the military education and training of members of the Armed Forces; the Services retain responsibility for managing the quality and content of their Services' PME programs at all levels within the guidelines of the military educational continuum and where appropriate, implementing policies contained in this document. T2 efforts and military education will remain coordinated and consistent. T2 decisions, initiatives or programs affecting military education will Enclosure A become operative when they have been reviewed and approved by the affected Services and the CJCS.

Appendix A to Enclosure A
Officer Professional Military Educational Continuum

1. Overview. The Officer PME Continuum (see Annex A to this Appendix) reflects the dynamic system of officer career education. It identifies areas of emphasis at each educational level and provides joint curriculum guidance for PME institutions. It is a comprehensive frame of reference depicting the progressive nature of PME and JPME, guiding an officer's individual development over time.

   a. The continuum structures the development of Service and joint officers by organizing the PME continuum into five military educational levels: precommissioning, primary, intermediate, senior, and G/FO. It defines the focus of each educational level in terms of the major levels of war (tactical, operational, and strategic) and links the educational levels so each builds upon the knowledge and values gained in previous levels.

   b. The continuum also recognizes both the distinctiveness and interdependence of joint and Service schools in officer education. Service schools, in keeping with their role of developing Service specialists, place emphasis on education primarily from a Service perspective in accordance with joint learning areas and objectives. Joint schools emphasize joint education from a joint perspective.

2. PME Relationships

   a. PME conveys the broad body of knowledge and develops the habits of mind essential to the military professional's expertise in the art and science of war. The PME system should produce:

      (1) Officers educated in the profession of arms who possess an intuitive approach to joint warfighting built upon individual Service competencies. Its aim is to produce graduates prepared to operate at appropriate levels of war in a joint environment and capable of generating quality tactical, operational, and strategic thought from a joint perspective.

    (2) Critical thinkers who view military affairs in the broadest context and are capable of identifying and evaluating likely changes and associated responses affecting the employment of US military forces.

    (3) Senior officers who can develop and execute national military strategies that effectively employ the Armed Forces in concert with other instruments of national power to achieve the goals of national security strategy and policy.

  b. JPME is that portion of PME that supports fulfillment of the educational requirements for joint officer management. Joint education prepares leaders to both conduct operations as a coherently joint force and to think their way through uncertainty.

3. The PME Continuum

  a. PME Levels. The continuum relates five military educational levels to five significant phases in an officer's career.

    (1) Precommissioning. Military education received at institutions and through programs producing commissioned officers upon graduation.

    (2) Primary. Education typically received at grades O-1 through O-3.

    (3) Intermediate. Education typically received at grade O-4.

    (4) Senior. Education typically received at grades O-5 or O-6.

    (5) General/Flag Officer. Education received as a G/FO.

  b. Levels of War. The continuum portrays the focus of each educational level in relation to the tactical, operational, and strategic levels of war as outlined in CJCS Manual 3500.04C, "Universal Joint Task List (UJTL)." It recognizes that PME and JPME curricula educate across levels of war.

  c. Precommissioning Education

    (1) Institutions and Courses

      (a) Military Service Academies.

      (b) Reserve Officer Training Corps (ROTC) units.

      (c) Federal and State Officer Candidate Schools (OCS) and Officer Training Schools (OTS).

    (2) Focus. Precommissioning education focuses on preparing officer candidates to become commissioned officers within the Military Department that administers the precommissioning program. The curricula are oriented toward providing candidates with a basic grounding in the US defense establishment and their chosen Military Service, as well as a foundation in leadership, management, ethics, and other subjects necessary to prepare them to serve as commissioned officers.

  d. Primary Education

    (1) Institutions and Courses

      (a) Branch, warfare or staff specialty schools.

      (b) Primary PME courses.

    (2) Focus. Primary education focuses on preparing junior officers to serve in their assigned branch or warfare or staff specialty. The curricula are predominantly Service oriented, primarily addressing the tactical level of war. Service schools that have programs centered on pay grade O-3 officers will foster an understanding of joint warfighting necessary for success at this level. Joint learning areas are embedded in Service PME instruction.

e. Intermediate Education
   (1) Institutions and Courses
       (a) Service Intermediate PME Institutions.
           1. Air Command and Staff College (ACSC).
           2. Army Command and General Staff College (ACGSC).
           3. College of Naval Command and Staff (CNCS) at the Naval War College.
           4. Marine Corps Command and Staff College (MCCSC).
           5. Service-recognized equivalent fellowships, advanced military schools, and international military colleges.
       (b) Joint Intermediate JPME Institutions.
           1. Joint and Combined Warfighting School (JCWS) at the Joint Forces Staff College (JFSC).
           2. Joint Advanced Warfighting School (JAWS) at the JFSC.
   (2) Focus. Intermediate education focuses on warfighting within the context of operational art. Students expand their understanding of joint force deployment and employment at the operational and tactical levels of war. They gain a better understanding of joint and Service perspectives. Inherent in this level is development of an officer's analytic capabilities and creative thought processes. In addition to continuing development of their joint warfighting expertise, they are introduced to theater strategy and plans, national military strategy, and national security strategy and policy.
f. Senior Education
   (1) Institutions and Courses
       (a) Service Senior PME Institutions.
           1. Air War College (AWC).
           2. Army War College (USAWC).
           3. College of Naval Warfare (CNW) at the Naval War College.
           4. Marine Corps War College (MCWAR).
           5. Service-recognized equivalent fellowships, advanced military schools and international military colleges.
       (b) Joint Senior JPME Institutions.
           1. National War College (NWC).
           2. Industrial College of the Armed Forces (ICAF).
           3. Joint and Combined Warfighting School (JCWS) at JFSC.
           4. Joint Advanced Warfighting School (JAWS) at JFSC.
   (2) Focus. To prepare students for positions of strategic leadership, senior education focuses on strategy, theater campaign planning, the art and science of developing, integrating and applying the instruments of national power (diplomatic, informational, military and economic) during peace and war. Studies at these colleges should emphasize analysis, foster critical examination, encourage creativity and provide a progressively broader educational experience.
g. Education for Reserve Component (RC) Officers. While RC officers participate in all of the previous PME and JPME levels, opportunities are limited for their attendance at JPME II. Accordingly, JFSC established the RC JPME program. This program contains a course similar in content, but not identical to, the in-residence JFSC course

for active component officers (O-4 to O-6). Phase I JPME is a prerequisite for this course per DODI 1215.20.

(1) Institution and Course. Advanced JPME (AJPME) Course at JFSC.

(2) Focus. Educates RC officers in joint operational-level planning and warfighting in order to instill a commitment to joint, interagency, and multinational teamwork, attitudes and perspectives.

h. G/FO education.

  (1) Institutions and Courses.

    (a) Joint G/FO PME programs.

      1. CAPSTONE course at NDU.

      2. Functional Component Commander Courses. Existing or potential functional component commander's courses, which are delivered by the Services, are valuable venues that serve both the educational and training needs of G/FOs. Services conducting these courses are encouraged to regularly review their curricula with the USJFCOM/J-7 to ensure currency and synergy with USJFCOM Joint Task Force (JTF) training efforts.

      3. Joint Flag Officer's Warfighting Course (JFOWC) at Air University.

      4. PINNACLE course at NDU.

  (2) Focus. Courses within the G/FO level of the JPME continuum prepare senior officers of the US Armed Forces for high-level joint, interagency and multinational responsibilities. Courses may address grand strategy, national security strategy, national military strategy, theater strategy and the conduct of campaigns and military operations in a joint, interagency and multinational environment to achieve US national interests and objectives. G/FO JPME is tiered to ensure the progressive and continuous development of executive level officers.

4. JPME Within the PME Continuum. Officer professional development and progression through the PME continuum is a Service responsibility. Embedded within the PME system, however, is a program of JPME overseen by the Joint Staff and designed to fulfill the educational requirements for joint officer management as mandated by the Goldwater-Nichols Act (GNA) of 1986. This JPME program comprises curriculum components in all five levels of the JPME continuum designed to develop progressively the knowledge, analytical skills, perspectives and values essential for US officers to function effectively in joint, interagency and multinational operations.

a. JPME Continuum and Flow. JPME includes five levels:

  (1) Preparatory JPME taught during precommissioning and primary schools.

  (2) JPME Phase I taught at Service intermediate-level colleges (ILC) and Service senior-level colleges (SLC) in-residence (for programs that have not been accredited for JPME II) or as a Distance Education (DE) or Distance Learning (DL) option.

  (3) JPME Phase II taught at the Joint Forces Staff College and Service SLCs.

  (4) The separate single-phase JPME programs at the National War College (NWC), Industrial College of the Armed Forces (ICAF) and Joint Advanced Warfighting School (JAWS).

  (5) G/FO courses.

b. All officers should complete precommissioning, primary and intermediate JPME. Officers striving for joint qualification shall complete JPME Phase I at ILC or the Service SLC; Phase II at either the resident SLCs (once accredited); JCWS; or the single-phase JPME programs at ICAF, NWC or JAWS. Officers selected for promotion to G/FO must attend and complete CAPSTONE within approximately 2 years after confirmation of selection to O-7 unless such attendance is waived per DODI 1300.20 (enclosure 8, paragraph E8.6). Finally, select G/FOs participate in JFOWC, the Functional Component Commander Courses and PINNACLE.

  c. JPME Emphasis in PME:

    (1) Precommissioning. In addition to an introduction to their respective Service, students should have knowledge of the basic US defense structure, roles and missions of other Military Services, the combatant command structure and the nature of American military power and joint warfare. (Appendix B to Enclosure E identifies joint learning areas and objectives for precommissioning-level programs.)

    (2) Primary (O-1 to O-3). JPME prepares officers for service in Joint Task Forces (JTF) where a thorough introductory grounding in joint warfighting is required. The programs at this level address the fundamentals of joint warfare, JTF organization and the combatant command structure, the characteristics of a joint campaign, how national and joint systems support tactical-level operations and the capabilities of the relevant systems of the other Services. (Appendix B to Enclosure E identifies joint learning areas and objectives for primary-level programs.)

    (3) Intermediate (O-4)

      (a) JPME Phase I (Service Colleges). Service ILCs teach joint operations from the standpoint of Service forces in a joint force supported by Service component commands. (Appendix C to Enclosure E identifies joint learning areas and objectives for Service intermediate programs.)

      (b) JPME Phase II. The Joint and Combined Warfighting School at JFSC examines joint operations from the standpoint of the CJCS, the JCS, a combatant commander and a JTF commander. It further develops joint attitudes and perspectives, exposes officers to and increases their understanding of Service cultures while concentrating on joint staff operations. (Appendix H to Enclosure E identifies joint learning areas and objectives for JPME Phase II.)

      (c) JAWS. Provides a separate, single-phase JPME curricula reflecting the distinct educational focus and joint character of its mission. Designed for a small group of selected Service-proficient officers (O-4 to O-6) enroute to planning related positions on the Joint Staff and in the combatant commands. The school's mission is to produce graduates that can create campaign-quality concepts, employ military power in concert with the other instruments of national power, accelerate transformation, succeed as joint force operational/strategic planners and commanders and be creative, conceptual, adaptive and innovative. The Services may recognize JAWS as an intermediate-level or senior-level PME equivalent. JAWS meets policies applicable to intermediate- and senior-level programs. (Appendix I to Enclosure E identifies joint learning areas and objectives for JAWS).

    (4) Senior (O-5 to O-6)

      (a) JPME Phase I and II (Service Colleges). Service SLCs provide JPME Phase I and in-resident JPME Phase II education. Service SLCs address theater- and national-level strategies and processes. Curricula focus on how the unified commanders, Joint Staff and DOD use the instruments of national power to develop and carry out national military strategy, develop joint operational expertise and perspectives and hone joint warfighting skills. (Appendix D and E to Enclosure E identify joint learning areas and objectives for Service senior-level programs.)

(b) JPME Phase II. JCWS at JFSC provides JPME Phase II for graduates of JPME Phase I programs to further develop joint attitudes and perspectives, joint operational expertise and hone joint warfighting skills. (Appendix H to Enclosure E identifies joint learning objectives for JPME Phase II.)

(c) JAWS. JAWS provides a separate single-phase JPME curricula reflecting the distinct educational focus and joint character of its mission. JAWS is designed for a small group of selected Service-proficient officers (O-4 to O-6) enroute to planning related positions on the Joint Staff and in the combatant commands. The school's mission is to produce graduates that can create campaign-quality concepts, employ military power in concert with the other instruments of national power, accelerate transformation, succeed as joint force operational/strategic planners and commanders and be creative, conceptual, adaptive and innovative. The Services may recognize JAWS as an intermediate-level or senior-level PME equivalent. JAWS meets policies applicable to intermediate- and senior-level programs (in a manner similar to other NDU colleges). (Appendix I to Enclosure E identifies joint learning objectives for JAWS).

(d) NWC. NWC provides a separate, single-phase JPME curricula reflecting the distinct educational focus and joint character of its mission. NWC's JPME curriculum focuses on national security strategy—the art and science of developing, applying and coordinating the instruments of national power (diplomatic, informational, military and economic,) to achieve objectives contributing to national security. (Appendix F to Enclosure E identifies joint learning areas and objectives for NWC.)

(e) ICAF. ICAF provides separate, single-phase JPME curricula reflecting the distinct educational focus and joint character of its mission. The ICAF JPME curriculum focuses on the resource component of national power, national resources and its integration into development and execution of national security strategy. (Appendix G to Enclosure E identifies joint learning areas and objectives for ICAF.)

(f) Advanced Joint Professional Military Education (AJPME). AJPME builds on the foundation established by the institutions teaching JPME Phase I. The course expands knowledge through hands-on learning and emphasizes national security systems, command structures, military capabilities, campaign planning and the integration of national resources. (Appendix J to Enclosure E identifies joint learning areas and objectives for AJPME).

(5) G/FO. G/FO JPME prepares senior officers of the US Armed Forces for high-level joint, interagency and multinational responsibilities. Courses may address grand strategy, national security strategy, national military strategy, theater strategy and the conduct of operational campaign in a joint, interagency and multinational environment to achieve US national objectives. (Appendices K-N to Enclosure E identify joint learning areas and objectives for G/FO JPME.)

Annex A
Appendix A
Enclosure A
GRADE
CADET/MISHIPMAN
O-1/O-2/O-3
O-4
O-5/O-6
O-7/O-8/O-9
EDUCATION LEVEL
PRECOMMISSIONING
PRIMARY
INTERMEDIATE
SENIOR
GENERAL/FLAG
EDUCATIONAL
INSTITUTIONS AND
COURSES
Service Academies
ROTC
OCS/OTS

- Branch, Warfare or Staff Specialty Schools
- Primary-Level PME Courses
- Air Command and Staff College
- Army Command and General Staff School
- College of Naval Command and Staff
- Marine Corps Command and Staff College
- JFSC, Joint and Combined Warfighting School
- JFSC, Joint Advanced Warfighting School1
- Air War College
- Army War College
- College of Naval Warfare
- Marine Corps War College
- Industrial College of the Armed Forces1
- National War College1
- JFSC, Joint and Combined Warfighting School
- JFSC, Joint Advanced Warfighting School1
- CAPSTONE
- Joint Functional Component Commander Courses
- Joint Flag Officer Warfighting Course
- PINNACLE

LEVELS OF WAR EMPHASIZED

Conceptual Awareness of all Levels

FOCUS OF MILITARY EDUCATION

Introduction to Services Missions

- Assigned Branch, Warfare or Staff Specialty
- Warfighting within the context of Operational Art
- Intro to theater strategy and plans, national military strategy and national security strategy
- Develop analytical capabilities and creative thought
- Service Schools: strategic leadership, national military strategy and theater strategy
- NWC: national security strategy
- ICAF: national security strategy with emphasis on the resource components
- Joint matters and national security
- Interagency process
- Multinational operations

JPME Phase I

- National military strategy
- National military capabilities command structure and strategic guidance
- Joint doctrine and concepts
- Joint and multinational forces at the operational level of war
- Joint planning and execution processes
- Information operations, C2 and battlespace awareness
- Joint force and joint requirements development

JPME Phase I

- National security strategy
- National planning systems and processes
- National and theater military strategy, campaigning and organization
- Joint doctrine, force and requirements development
- Information operations, C2 and battlespace awareness
- Joint strategic leader development

JPME Phase II

- National security strategy
- National military strategy and organization
- Joint warfare, theater strategy and campaigning
- National and joint planning systems and processes
- Integration of Joint, IA and multinational capabilities
- Information ops, C2 and battlespace awareness
- Joint force and joint requirements development
- Joint strategic leader development

JOINT EMPHASIS

Joint Introduction

- National Military Capabilities and Organization
- Foundation of Joint Warfare

Joint Awareness

- Joint Warfare Fundamentals
- Joint Campaigning

AJPME and JPME Phase II

- National strategic security systems and guidance and command structures
- Theater strategy and campaigning
- Integration of Joint interagency (IA) and multinational capabilities
- Information operations
- Joint planning systems

CAPSTONE

- National security strategy
- Joint operational art

Joint Functional Component Commander Courses & JFOWC

- National security strategy
- National planning systems and organization
- National military strategy & organization
- Theater strategy, campaigning and military operations in Joint, interagency, and multi-national environment
- Information operations
- Strategic leader development

PINNACLE

- Joint/Combined force development
- Building & commanding the joint combined force
- The JFC and the IA, NCA, NMS and the Congress

TACTICAL

1ICAF, NWC, and JAWS offer single-phase JPME

OPERATIONAL STRATEGIC

Joint Officer Management Educational Requirements

1. General
   a. This appendix provides guidance for the Military Services concerning statutory educational requirements based1 on title 10, USC, chapter 107. Additional guidance concerning joint officer management can be found in DODI 1300-20, "DOD Joint Officer Management Program Procedures" (reference e) and DODI 1215.20, "Reserve Component (RC) Joint Officer Management Program" (reference d).
   b. Within the DOD Joint Officer Management Program, a selected officer with the educational and joint duty prerequisites may be designated as "joint specialty officer (JSO)" or "JSO nominee," an administrative classification that identifies an officer as having education and/or experience in joint matters.
   c. The Reserve Component (RC) Joint Officer Management Program addresses management of RC officers on the Reserve Active Status List (RASL).
2. Educational Requirements for Joint Specialty Officers. To satisfy the educational prerequisites for JSO/JSO nominee designation, officers must receive credit for completing a CJCS-certified or accredited program of JPME. In exceptional cases, CJCS may grant JPME credit to officers who have not completed the full course of study. AJPME as a JPME analog does not satisfy the educational prerequisites for JSO/JSO nominee designation. Paths for satisfying the educational requirements for JSO/JSO nominee designation can be accomplished in several ways:
   a. An officer completes JPME Phase I at a Service ILC or SLC. This is followed by completion of JPME Phase II at JCWS or an accredited Service SLC. Other than officers possessing a critical occupational specialty, officers must attend JPME II prior to completion of the joint assignment to qualify for JSO designation. The SecDef can waive this requirement for a limited number of officers designated as JSOs in a fiscal year. The following additional conditions apply:
      1. Formerly found under title 10, USC, chapter 38, section 663.
         (1)Attendance at JPME Phase II prior to completion of JPME Phase I requires approval of a Direct Entry Waiver by the CJCS. Such waiver requests must be submitted in writing by the officer's Service to the Joint Staff/J-1 a minimum of 60 days prior to the start of the JCWS class to which the Service desires to send the officer.
         (2)Waivers are to be held to a minimum, with approval granted on a case-by-case basis for compelling reasons. Waiver requests require justification and must demonstrate critical career timing precluding the officer from attending JPME Phase I prior to Phase II. Requests must address the officer's qualifications, JSO potential and plans for subsequent assignment to a JDA. Waiver approval must be received prior to attendance at JCWS. Waiver approval is for the sequencing of JPME phases only and does not remove the JSO educational requirement to complete JPME Phase I.
         (3)Officers granted direct-entry waivers will be scheduled to attend the 5-day Joint Transition Course conducted by the JFSC immediately prior to beginning their Phase II course.
   b. An officer completes an intermediate- or senior-level international military education program for which JPME Phase I equivalent credit has been approved by the

CJCS. (This method for receiving JPME Phase I credit is subject to the provisions of paragraph 4 of this appendix.) This is followed by completion of JPME Phase II at JCWS or in-resident attendance at an accredited Service SLC.

   c. An officer completes NWC, ICAF or JAWS when accredited.

3. Educational Requirements for Joint Duty Assignments Reserve (JDAR). To the extent practical, Reserve officers on the DOD RASL will complete the appropriate level of educational requirements before assignment to a JDA-R billet. Positions will be validated and documented to identify positions that require no JPME, those that require JPME Phase I and those that require AJPME. Officers in critical JDA-R billets will complete AJPME before assignment, where practicable. Additional guidance concerning Reserve officer joint officer management can be found in DODI 1215.20.

4. Equivalent JPME Phase I Credit. The CJCS authorizes the Service Chiefs to award JPME Phase I credit to officers who successfully complete a resident international military college, subject to the provisions cited below.

   a. The resident international military college is on the CJCS approved JPME Phase I Equivalency list.

   b. Individuals selected for these programs meet the same rigorous selection criteria as other ILC and SLC PME attendees.

   c. The Service grants PME credit for completion of the international military college programs.

5. CJCS Accredited JPME Programs. The Chairman accredits JPME programs at all ILCs and SLCs under the provisions of the PAJE (Enclosure F) . . .

ENCLOSURE B
POLICIES FOR INTERMEDIATE- AND SENIOR-LEVEL COLLEGES

1. General. This enclosure outlines policies applicable to intermediate and senior PME programs.

2. International Officer Participation. The Services and NDU may maintain international officer programs that best meet their respective colleges' missions. International officer participation will be consistent with relevant security considerations and appropriate directives.

3. Civilian Participation. The Services and NDU may include civilian students in their programs. Civilian students should have appropriate academic and professional backgrounds. Participation by both DOD and non-DOD civilian students is desired, with focus of non-DOD students on perspectives of the interagency.

4. Curricula. PME institutions will base their curricula on their parent Service's needs or, in the case of the NDU colleges, on their CJCS assigned missions. JPME I and II will not be delivered as a stand-alone course, they must be delivered in conjunction with Service PME. Each college will fulfill the appropriate joint learning objectives and generally have a curriculum that includes:

   a. Mission-specific courses appropriate to the Service or college.

   b. JPME conducted within the context of the college or school mission. (Enclosure E identifies the joint learning areas and objectives for intermediate and senior PME colleges and schools.)

   c. Elective courses that enhance each student's professional and educational opportunities.

5. Resident Programs
   a. Class and Seminar Mix
      (1) Class mix at each Service ILC and Service SLC will contain a balanced mix of operational and functional expertise from the two nonhost Military Departments. Service SLCs shall have no more than 60 percent host Military Department student representation across their student bodies. This percentage is computed by including US military officers, international officers and civilian enrollments in the student body.
      (2) Seminar mix at Service ILCs and Service SLCs must include at least one officer from each of the two non-host Military Departments.
      (3) NWC, ICAF and JAWS must have approximately equal representation from each of the three Military Departments in their military student bodies.
      (4) JFSC military student quotas in JCWS will be allocated in accordance with the distribution of billets by Service on the Joint Duty Assignment List (JDAL). AJPME quotas will have approximately equal representation from each of the three Military Departments.
      (5) For all intermediate- and senior-level schools, Navy, Marine Corps and Coast Guard officers will count toward Sea Service Student requirements.
   b. Faculty. Faculty members will be of the highest caliber, combining the requisite functional or operational expertise with teaching ability and appropriate academic credentials.
      (1) Military Faculty. Active duty military officers bring to a faculty invaluable operational currency and expertise; therefore, a sufficient portion of each college/school's faculty shall be active duty military officers. Military faculty are those uniformed personnel who prepare, design or teach PME curricula or conduct research related to PME. Navy, Marine Corps and Coast Guard officers count toward Sea Service military faculty requirements.
         (a) Faculty Mix. Personnel performing strictly administrative functions may not be counted in faculty ratios and mixes.
            1. Service SLCs. Total non-host Military Department faculty shall be no less than 40 percent of the total military faculty whose primary duty is student instruction of JPME. The mix of the faculty members should be proportionally divided among each non-host Military Department.
            2. Service ILCs. The mix of military faculty members whose primary duty is student instruction of JPME should be a minimum of 5 percent from each non-host Military Department.
            3. NDU. At NWC, ICAF and JFSC, the mix of military faculty members will be approximately one-third from each Military Department.
         (b) Qualifications
            1. Service SLCs. 75 percent of the military faculty should be graduates of a senior-level PME program or be JSOs.
            2. Service ILCs. 75 percent of the military faculty should be graduates of an intermediate- or senior-level PME program or be JSOs.

3. JFSC. All military faculty at JFSC should be graduates of an intermediate-
or senior-level PME program or have comparable joint experience.

(2) Civilian Faculty. The Services and NDU determine the appropriate number of
civilians on their respective college faculties. Civilian faculty members should
have strong academic records.

(3) Faculty Chairs

   (a) Each NDU JPME College will establish a CJCS Professor of Military Studies Chair. CJCS chairs will be military faculty of appropriate rank who have
   completed JPME (or are JSOs), have recent joint operational experience and
   are capable of contributing insight into joint matters to the faculty and student body. The CJCS approves nominees for these chairs, which will be
   filled from authorized military faculty positions. CJCS chairs act as a direct
   liaison with the Office of the CJCS and the Joint Staff.

   (b) Each NDU JPME College is encouraged to establish similar Service Chiefs
   chairs' for each of the Services.

   (c) Each Service College is encouraged, within its own resources, to establish
   CJCS chairs as described above, as well as similar Service Chiefs' chairs for
   each non-host Service.

(4) Student-to-Faculty Ratios

   (a) Reasonable student-to-faculty ratios are essential to quality instruction. The
   following ratios are standards for the PME level indicated:
   1. ILC/JCWS – 4:1.
   2. SLC/JAWS – 3.5:1.

   (b) These ratios are computed by dividing the total number of students by the
   total faculty using the following guidelines:
   1. Faculty. Personnel (military and civilian) who—as determined by the
   college or school—teach, prepare or design PME curriculum, or conduct research related to PME, count in computation of this ratio. Personnel performing strictly administrative functions may not be counted
   as faculty for computing student-to-faculty ratios.
   2. Students. All (US and international) military officers and civilians assigned to the institution as students for the purpose of completing a
   prescribed course of instruction count as students in the computation of
   student-to-faculty ratios. Non-host Military Departments must provide
   ILC and SLC students who reflect a representative mix of operational
   and functional expertise from that Department.

c. Learning Methodology. PME institutions will primarily use a mix of active learning methods such as research, writing, reading, oral presentations, seminar discussions, case studies, wargaming, simulations and distributive learning. Passive
learning methods (without student interaction) may also be used to enhance the
overall educational experience. Small group learning should be the principal resident education methodology.

6. Non-Resident Education Programs

   a. Non-resident programs offer the opportunity to provide PME and JPME to a larger
   population than can be supported in resident facilities. These programs must be
   of sufficient substance and academic rigor—measured against challenging, realistic

standards—that they clearly achieve the objectives of this instruction. Such educational standards must accommodate the differences in the non-resident environments, non-resident methodologies and needs of non-resident students.

b. Non-resident education is the delivery of a structured curriculum to a student available at a different time or place than the teaching institution's resident program. It is a deliberate and planned learning experience that incorporates both teaching by the sponsoring institution as well as learning efforts by the student. Non-resident education provides instruction in places or times that are convenient and accessible for learners rather than teachers or teaching institutions. To accomplish this, the educational institution uses special course design, instructional techniques, methods of communication and contact with students and organizational and administrative arrangements to create a quality learning experience. There are three approaches used to provide nonresident JPME via an appropriate, structured curriculum.

(1) Satellite seminars or classes. The satellite approach typically uses adjunct faculty to replicate the in-residence learning experience at a location away from the JPME institution. The instructional format is essentially the same as that provided to in-residence students.

(2) Distance/Distributed Learning (DL). In a DL format there is a separation of either time or distance between the instructor and the learner or learners. JPME via DL can be designed to serve individual learners or distributed virtual seminars of learners. It typically employs combinations of print or electronic media, combined with appropriate technologies such as Video Tele-Education (VTE) and web-based applications. The web-based formats may also be combinations of either asynchronous (self-paced / at different times) or synchronous (real-time interaction) delivery strategies.

(3) Blended learning. A blended approach combines DL with some form of in-residence program. The in-residence phase or phases are typically at the JPME institution, but can be conducted at satellite facilities.

c. JPME Learning Objectives. Non-resident programs must meet the JPME learning objectives assigned to their respective resident institutions. Non-resident curricula and related educational products and materials should derive from and closely parallel the Program of Instruction (POI)/curriculum of their respective resident institutions. The differences between the two types of programs are primarily in the specific delivery methodology and techniques employed to achieve the PME and JPME learning objectives.

d. Class and Seminar Mix. With the exception of AJPME, nonresident programs need not maintain the mix of students by Service in their overall student bodies and seminars required of resident programs. ILC and SLC non-resident programs should, when delivered in a group environment, seek diversity in student populations by providing enrollment opportunities to non-host Services, Reserve Components, DOD and non-DOD civilians, as appropriate.

e. Faculty

(1) Qualifications. Non-resident program faculty will meet the same qualification criteria as faculty in their respective resident institutions.

(2) Faculty Mix. With the exception of AJPME, non-resident programs do not require the same faculty mix as resident programs and specific percentages do not apply. Service ILCs and SLCs must show that non-host Service faculty members

are an integral part of the development and implementation of their non-resident curriculum.

   f. Student-Faculty Ratios

      (1) In non-resident education programs, the number of faculty members is determined by the course design and the demands of students—what the methodology requires and how much access students need to faculty to successfully master the subject matter. Service ILCs, SLCs and JFSC must show proper faculty staffing for the methodology being used and that all students have reasonable access to faculty subject matter expertise and counseling.

      (2) In determining appropriate non-resident faculty staffing levels, institutions should consider all faculty actively participating in the development and implementation of the program.

   g. Learning Methodology

      (1) Service ILCs and SLCs may choose methodologies and techniques appropriate to their Service, subject content and student populations.

      (2) Non-resident programs must demonstrate they provide their students with an understanding of other Services' perspectives in building a joint perspective. Service ILCs, SLCs and JFSC must show they have a valid non-resident methodology for developing joint perspective and must demonstrate through evaluation of student performance and outcomes assessment that students are acquiring the desired joint perspective . . .

ENCLOSURE E

JOINT PROFESSIONAL MILITARY EDUCATION

1. General. This enclosure provides common educational standards, taxonomy of desired levels of learning achievement and joint learning objectives for the five levels of PME.

2. Common Educational Standards. The following describes educational standards common to all PME institutions that the CJCS considers essential for satisfactory resident and non-resident programs. Each standard is described primarily in qualitative terms, since no particular organizational pattern or application strategy applies in all settings.

   a. Standard 1—*Develop Joint Awareness, Perspective, and Attitudes.* JPME curricula should prepare graduates to operate in a joint, interagency and multinational environment and bring a joint perspective to bear in their tactical, operational, strategic and critical thinking as well as professional actions. Institutions' missions, goals, objectives, educational activities and the mix of students and faculty should reflect joint educational requirements, encourage critical analyses of current and emerging national strategies from a joint perspective and foster a commitment to joint and interagency cooperation. The institutions' leadership, faculty, and students should manifest an appropriate commitment to jointness.

   b. Standard 2—*Employ Predominately Active and Highly Effective Instructional Methods.* Instructional methods should be appropriate to the subject matter and desired level of learning and should employ active student learning whenever feasible. The goals of the educational offerings are rigorous and challenging, requiring students to engage in critical thinking and active interchange with faculty and students.

   c. Standard 3—*Assess Student Achievement.* Each institution should aggressively assess its students' performance. Educational goals and objectives should be clearly stated and students' performance should be measured against defined institutional standards by appropriate assessment tools to identify whether desired educational outcomes are being achieved.

d. Standard 4—*Assess Program Effectiveness.* Institutions should conduct surveys of students, graduates, their supervisors and the joint 7 Supervisor surveys are optional for non-resident programs. Leadership to determine curricula and educational effectiveness of their academic programs. Additionally, institutions should analyze student performance for indicators of program effectiveness. Results of these analyses should be used to refine or develop curricula that continue to meet evolving mission requirements in the context of an ever-changing world. Curricula should be the product of a regular, rigorous and documented review process.

e. Standard 5—*Conduct Quality Faculty Recruitment. Selection, Assignment and Performance Assessment Program.* Faculty should have the academic credentials, teaching skills and experience in joint and professional matters needed to teach in the institution. Faculty roles and responsibilities should be clearly documented. Institutions should hold faculty accountable to clearly defined and measurable performance criteria and standards.

f. Standard 6—*Conduct Faculty Development Programs For Improving Instructional Skills and Increasing Subject Matter Mastery.* Each institution should have a faculty development program to refine teaching skills, improve instructional methods, maintain currency in subject areas and encourage further professional development. Policy and resources must support the faculty development program.

g. Standard 7—*Provide Institutional Resources to Support the Educational Process.* Each institution must have a library or learning resource center, informational resources, financial resources and physical resources that meet the needs of all users and supports the mission and programs of the institution.

3. Levels of Learning Achievement. See Appendix A to Enclosure E.

LEARNING OBJECTIVE VERBS

Levels of Learning Achievement. Below is a list of descriptive verbs that constitute a useful hierarchy of possible levels of learning. The verbs, listed in increasing levels of achievement, are used to define the JPME objectives in the following appendixes of Enclosure E.

Level Illustrative Level Definitions

- Knowledge Arrange, define, describe, identify, know, label, list, match, memorize, name, order, outline, recognize, relate, recall, repeat, reproduce, select, state
- Remembering previously learned information
- Comprehension Classify, comprehend, convert, define, discuss, distinguish, estimate, explain, express, extend, generalize, give example(s), identify, indicate, infer, locate, paraphrase, predict, recognize, rewrite, report, restate, review, select, summarize, translate
- Grasping the meaning of information

- Value Accepts, adopts, approves, completes, chooses, commits, demonstrates, describes, desires, differentiates, displays, endorses, exhibits, explains, expresses, follows, forms, initiates, invites, joins, justifies, prefers, proposes, reads, reports, sanctions, selects, shares, studies, value, works Internalization and the consistent display of a behavior. The levels of valuing consist of acceptance of a value, preference for a value and commitment (conviction)
- Application Apply, change, choose, compute, demonstrate, discover, dramatize, employ, illustrate, interpret, manipulate, modify, operate, practice, predict, prepare, produce, relate, schedule, show, sketch, solve, use, write
- Applying knowledge to actual situations
- Analysis Analyze, appraise, breakdown, calculate, categorize, classify, compare, contrast, criticize, derive, diagram, differentiate, discriminate, distinguish, examine, experiment, identify, illustrate, infer, interpret, model, outline, point out, question, related, select, separate, subdivide, test Breaking down objects or ideas into simpler parts and seeing how the parts relate and are organized
- Synthesis Arrange, assemble, categorize, collect, combine, comply, compose, construct, create, design, develop, devise, explain, formulate, generate, plan, prepare, propose, rearrange, reconstruct, relate, reorganize, revise, rewrite, set up, summarize, synthesize, tell, write
- Rearranging component ideas into a new whole
- Evaluation Appraise, argue, assess, attach, choose, compare, conclude, contrast, defend, describe, discriminate, estimate, evaluate, explain, judge, justify, interpret, relate, predict, rate, select, summarize, support, value
- Making judgments based on internal evidence or external criteria

Appendix B to Enclosure E
Precommissioning and Primary Joint Professional
Military Education

1. Precommissioning
   a. Institutions and Programs
      (1) Military Service Academies.
      (2) ROTC units.
      (3) OCS and OTS.
   b. Joint Emphasis. In addition to an introduction to their respective Service, students should have knowledge of the basic US defense structure, roles and missions of other Military Services, the combatant command structure and the nature of American military power and joint warfare.
   c. Learning Area 1—*National Military Capabilities and Organization*
      (1) Know the organization for national security and how defense organizations fit into the overall structure.
      (2) Know the organization, role and functions of the JCS.
      (3) Know the chain of command from the President and the SecDef to the individual Service headquarters and to the unified commands.
      (4) Know the primary missions and responsibilities of the combatant commands.
      (5) Know the Military Services' primary roles, missions and organizations.

   d. Learning Area 2—*Foundation of Joint Warfare*
     (1) Describe the nature of American Military Power (Chapter 1, Joint Pub 1—reference l).
     (2) Identify the values in Joint Warfare (Chapter 2, Joint Pub 1).
     (3) Understand fundamentals of information operations.
     (4) Know how to access joint learning resources.

2. Primary
   a. Institutions and Courses
     (1) Branch, warfare and staff specialty schools.
     (2) Primary PME courses.
   b. Joint Emphasis. Prepares officers for service in Joint Task Forces (JTF) where a thorough introduction in joint warfighting is required. The programs at this level address the fundamentals of joint warfare, JTF organization and the combatant command structure, the characteristics of a joint campaign, how national and joint systems support tactical-level operations and the capabilities of the relevant systems of the other Services.
   c. Learning Area 1—*Joint Warfare Fundamentals*
     (1) Know fundamentals of joint warfare (Chapter 3, Joint Pub 1).
     (2) Know each combatant command's mission, organization and responsibilities.
     (3) Comprehend joint aspects of Stability Operations.
     (4) Comprehend, within the context of the prevailing national military strategic focus, how national and joint systems are integrated to support Service tactical planning and operations (for tactical battlespace being taught at school).
     (5) Know the capabilities of other Services' weapon systems pertinent to the Service host-school systems and the synergistic effect gained from effective use of their joint capabilities.
     (6) Comprehend the effects that can be achieved with information operations and the implications for tactical operations.
     (7) Know how to access joint learning resources.
   d. Learning Area 2—*Joint Campaigning*
     (1) Know who can form a JTF and how and when a JTF is formed.
     (2) Know the fundamentals of a JTF organization.
     (3) Comprehend the characteristics of a joint campaign and the relationships of supporting capabilities (Chapter 4, Joint Pub 1).
     (4) Recognize the roles that factors such as geopolitics, culture and religion play in shaping planning and execution of joint force operations...

Appendix C TO Enclosure E
Service Intermediate-Level College (ILC) Joint Learning
Areas and Objectives

1. Overview. The Service ILCs' curricula focus is warfighting within the context of operational art.
2. Mission. The Service ILCs' joint mission is to expand student understanding, from a Service component perspective, of joint force employment at the operational and tactical levels of war.

3. Learning Area 1—*National Military Capabilities, Command Structure and Strategic Guidance*
   a. Comprehend the capabilities and limitations of US military forces to conduct the full range of military operations against the capabilities of 21st century adversaries.
   b. Comprehend the organizational framework within which joint forces are created, employed and sustained.
   c. Comprehend the purpose, roles, functions and relationships of the President and the SecDef, National Security Council (NSC), CJCS, JCS, combatant commanders, joint force commanders (JFCs), Service component commanders and combat support organizations.
   d. Comprehend how joint force command relationships and directive authority for logistics support joint warfighting capabilities.
   e. Comprehend how the US military is organized to plan, execute, sustain and train for joint, interagency and multinational operations.
   f. Comprehend the strategic guidance contained in the national security strategy, national military strategy and national military strategy for the global war on terrorism.
4. Learning Area 2—*Joint Doctrine and Concepts*
   a. Comprehend current joint doctrine.
   b. Comprehend the factors and emerging concepts influencing joint doctrine.
   c. Apply solutions to operational problems using current joint doctrine.
   d. Comprehend the interrelationship between Service doctrine and joint doctrine.
5. Learning Area 3—*Joint and Multinational Forces at the Operational Level of War*
   a. Comprehend the considerations for employing joint and multinational forces at the operational level of war.
   b. Comprehend how theory and principles of war pertain to the operational level of war.
   c. Analyze a plan for employment of joint forces at the operational level of war.
   d. Comprehend the relationships among national objectives, military objectives and conflict termination, as illustrated by previous wars, campaigns and operations.
   e. Comprehend the relationships among the strategic, operational and tactical levels of war.
   f. Comprehend the relationships between all elements of national power (diplomatic, informational, military and economic) and the importance of interagency and multinational coordination in these elements, including homeland security and defense.
6. Learning Area 4. *Joint Planning and Execution Processes*
   a. Comprehend the relationship among national objectives and means available through the framework provided by joint planning processes.
   b. Comprehend the effect of time, coordination, policy changes and political development on the planning process.
   c. Comprehend how the defense planning systems affect joint operational planning and force planning.
   d. Comprehend how national, joint and Service intelligence organizations support JFCs and their Service component commanders.
   e. Comprehend the fundamentals of campaign planning.
   f. Comprehend the roles that factors such as geopolitics, geostrategy, society, culture and religion play in shaping planning and execution of joint force operations across the range of military operations.

7. Learning Area 5—*Information Operations, Command and Control (C2) and Battlespace Awareness*
   a. Comprehend how information operations are integrated in support of national and military strategies.
   b. Comprehend how information operations are incorporated into both deliberate and crisis-action planning processes at the operational and JTF levels.
   c. Know how C2 and battlespace awareness apply at the operational level of war and how they support operations conducted by a networked force.
   d. Comprehend how increased reliance on information technology throughout the range of military operations creates opportunities and vulnerabilities.

Appendix D to Enclosure E
Service Senior-Level Colleges (SLC) Joint Learning Areas
and Objectives (JPME PHASE I)

1. Overview. Service SLCs focus on national military strategy as derived from national security strategy and policy, and its impact on strategic leadership, force readiness, theater strategy and campaigning.
2. Mission. Although each Service SLC mission is unique, a fundamental objective of each is to prepare future military and civilian leaders for high-level policy, command and staff responsibilities by educating them in the diplomatic, informational, military and economic dimensions of the strategic security environment and the effect of those dimensions on strategy formulation, implementation and campaigning. SLC subject matter is inherently joint; JPME at this level focuses on the development of joint attitudes and perspectives.
3. Learning Area 1—*National Security Strategy*
   a. Analyze the strategic art; i.e., developing, applying and coordinating the instruments of national power to secure national security objectives.
   b. Comprehend how national policy is turned into executable military strategies.
   c. Analyze how the constituent elements of government and American society exert influence on the national strategy process.
4. Learning Area 2—*National Planning Systems and Processes*
   a. Evaluate the DOD systems and processes by which national ends, ways and means are reconciled, integrated and applied.
   b. Analyze how time, coordination, policy, politics, doctrine and national power affect the planning process.
   c. Analyze and apply the principal joint strategy development and operational planning processes.
   d. Analyze the role of joint doctrine with respect to unified command.
   e. Analyze how the interagency's structure and processes influence the planning for and application of the military instrument of national power.
5. Learning Area 3—*National Military Strategy and Organization*
   a. Comprehend the art and science of developing, deploying, employing and sustaining the military resources of the Nation, in concert with other instruments of national power, to attain national security objectives.
   b. Evaluate the national military strategy, especially with respect to the changing nature of warfare.

   c. Analyze the roles, relationships, and functions of the President, SecDef, CJCS, JCS, combatant commanders, Secretaries of the Military Departments and the Service Chiefs.

   d. Evaluate how the capabilities and limitations of the US force structure affect the development of joint military strategy.

   e. Apply an analytical framework that incorporates the role that factors such as geopolitics, geostrategy, society, culture and religion play in shaping the desired outcomes of policies, strategies and campaigns in the joint, interagency, and multinational arena.

6. Learning Area 4—*Theater Strategy and Campaigning*

   a. Analyze how joint, unified and multinational campaigns and operations support national objectives and relate to the national strategic, theater strategic and operational levels of war.

   b. Synthesize the role and perspective of the combatant commander and staff in developing various theater policies, strategies and plans.

   c. Analyze joint operational art and emerging joint operational concepts.

   d. Appraise processes for coordinating US military plans and actions effectively with forces from other countries and with interagency and non-governmental organizations to include homeland security and defense.

7. Learning Area 5—*Information Operations, C2 and Battlespace Awareness*

   a. Analyze how information operations are integrated to support the national military and national security strategies and the interagency process.

   b. Analyze how information operations apply at the operational and strategic levels of war and how they support the operations of a networked force.

   c. Analyze the integration of information operations, C2 and battlespace awareness to theater campaign development.

   d. Analyze the principles, capabilities and limitations of information operations across the range of military operations and plans—to include pre- and post-conflict operations.

   e. Analyze the use of information operations to achieve desired effects across the spectrum of national security threats.

8. Learning Area 6—*Joint Strategic Leader Development*

   a. Synthesize techniques for leading in a joint, interagency and multinational environment.

   b. Synthesize leadership skills necessary to sustain innovative, agile and ethical organizations in a joint, interagency and multinational environment.

Appendix E to Enclosure E
Service Senior-Level Colleges (SLC) Joint Learning Areas
and Objectives (JPME PHASE II)

1. Overview. Service SLCs focus on national military strategy as derived from national security strategy and policy, and its impact on strategic leadership, force readiness, theater strategy and campaigning, and joint warfighting.

2. Mission. Although each Service SLC mission is unique, a fundamental objective of each is to prepare future military and civilian leaders for high-level policy, command and staff responsibilities requiring joint and Service operational expertise and warfighting

skills by educating them in the diplomatic, informational, military and economic dimensions of the strategic security environment and the effect of those dimensions on strategy formulation, implementation and campaigning. SLC subject matter is inherently joint; JPME at this level focuses on the immersion of students in a joint, interagency, and multinational environment and completes educational requirements for JSO nomination.

3. Learning Area 1—*National Security Strategy*
   a. Apply key strategic concepts, logic and analytical frameworks to the formulation and evaluation of strategy.
   b. Evaluate historical and/or contemporary applications of national security strategy to include the current US national security strategy and military strategy.
   c. Apply appropriate strategic security policies, strategies, and guidance used in developing plans across the range of military operations to support national objectives.
   d. Analyze the integration of all instruments of national power in achieving strategic objectives, with a focus on the employment of the military instrument of national power both as a supported instrument and as a supporting instrument of national power.

4. Learning Area 2—*National Military Strategy and Organization.*
   a. Comprehend the art and science of developing, deploying, employing and sustaining the military resources of the Nation, in conjunction with other instruments of national power, to attain national security objectives.
   b. Evaluate the national military strategy, especially with respect the changing nature of warfare.
   c. Analyze the roles, relationships, and functions of the President, SecDef, CJCS, Joint Staff, Combatant Commanders, Secretaries of the Military Departments and the Service Chiefs.
   d. Evaluate how the capabilities and limitations of the US force structure affect the development of joint military strategy.

5. Learning Area 3—*Joint Warfare, Theater Strategy and Campaigning.*
   a. Evaluate the principles of joint warfare, joint military doctrine and emerging concepts to joint, unified, interagency and multinational operations, in peace and war.
   b. Evaluate how joint, unified, and multinational campaigns and operations support national objectives and relate to the national strategic, national military strategic, theater strategic and operational levels in war.
   c. Synthesize how national military and joint theater strategies meet national strategic goals across the range of military operations.
   d. Synthesize the role and perspective of the combatant commander and staff in developing various theater policies, strategies, and plans to include WMD/E.
   e. Apply an analytical framework that incorporates the role that factors such as geopolitics, geostrategy, society, culture and religion play in shaping the desired outcomes of policies, strategies and campaigns in the joint, interagency, and multinational arena.

6. Learning Area 4—*National and Joint Planning Systems and Processes*
   a. Evaluate the DOD systems and processes by which national ends, ways and means are reconciled, integrated and applied.
   b. Analyze how time, coordination, policy, politics, doctrine and national power affect the planning process.

   c. Analyze and apply the principal joint strategy development and operational planning processes.

   d. Analyze the role of joint doctrine with respect to unified command.

   e. Analyze how the interagency structures and processes influence the planning for and application of the military instrument of national power.

7. Learning Area 5—*Integration of Joint. Interagency and Multinational Capabilities*

   a. Synthesize the capabilities and limitations of all Services (own Service, other Services—to include Special Operations Forces (SOF)) in achieving the appropriate strategic objectives in joint, interagency, and multinational operations.

   b. Analyze the capabilities and limitations of multinational forces in achieving the appropriate strategic objectives in coalition operations. c. Analyze the capabilities and limitations of the interagency processes in achieving the appropriate strategic objectives in joint plans.

   c. Analyze the integration of joint, interagency, and multinational capabilities across the range of military operations and plans - both in preparation and execution phases - and evaluate its success in achieving the desired effects.

   d. Comprehend the attributes of the future joint force and how this force will organize, plan, prepare and conduct operations.

   e. Value a thoroughly joint perspective and appreciate the increased power available to commanders through joint, combined, interagency efforts and teamwork.

8. Learning Area 6—*Information Operations, C2 and Battlespace Awareness.*

   a. Analyze how information operations are integrated to support the national military and national security strategies and the interagency process.

   b. Analyze how information operations apply at the operational and strategic levels of war and how they support the operations of a networked force.

   c. Analyze the integration of information operations, C2 and battlespace awareness to theater campaign development.

   d. Analyze the principles, capabilities and limitations of information operations across the range of military operations and plans—to include pre- and post-conflict operations.

   e. Analyze the use of information operations to achieve desired effects across the spectrum of national security threats.

9. Learning Area 7—*Joint Strategic Leader Development.*

   a. Synthesize techniques for leading in a joint, interagency and multinational environment.

   b. Synthesize leadership skills necessary to sustain innovative, agile and ethical organizations in a joint, interagency and multinational environment.

Appendix F to Enclosure E
National War College (NWC) Joint Learning
Areas and Objectives

1. Mission. The NWC mission is to educate future leaders of the Armed Forces, Department of State and other civilian agencies for high-level policy, command and staff responsibilities by conducting a senior-level course of study in national security strategy.

2. Focus. The NWC curriculum focuses on national security strategy. It provides graduate education in that subject to senior military and civilian leaders with an emphasis on

both the joint military and interagency dimensions of national security strategy. The NWC program concentrates on developing the habits of mind, conceptual foundations and critical faculties graduates will need at their highest level of strategic responsibility. Its goal is to produce national security practitioners who can develop and implement national security strategy holistically by orchestrating all the instruments of national power in a coherent plan to achieve national objectives in peace, crisis and war. NWC provides a distinct, single-phase JPME program tailored to its particular mission and focus that fully satisfies educational requirements for joint officer management.

3. Learning Area 1—*National Security Strategy*
   a. Analyze key concepts in national security strategy, their logical interrelationships, and analytical frameworks incorporating them.
   b. Apply key strategic concepts, logic and analytical frameworks to the formulation and evaluation of strategy.
   c. Evaluate historical and/or contemporary applications of national security strategy to include the current US national security strategy.
   d. Develop effective national security strategies for specific security challenges to include combating WMD/E, homeland security and defense and prepare national-level implementing guidance.
4. Learning Area 2—*Geo-Strategic Context*
   a. Comprehend the major social, cultural, political, economic, military, technological and historical issues in selected states and regions.
   b. Comprehend the roles and influence of international organizations and other non-state actors.
   c. Evaluate key military, non-military and transnational challenges to US national security.
   d. Conduct strategic assessments of selected international regions, states or issues from both US and selected "other actor" perspectives.
   e. Apply an analytical framework that incorporates the role that factors such as geopolitics, geostrategy, society, culture and religion play in shaping the desired outcomes of policies, strategies and campaigns in the joint, interagency, and multinational arena.
5. Learning Area 3—*Instruments of National Power*
   a. Comprehend the fundamental characteristics, capabilities and limitations of diplomatic, informational, military and economic instruments of national power.
   b. Investigate concepts and approaches for the employment of diplomatic, informational, military and economic instruments in support of national security strategy.
   c. Evaluate selected examples of the strategic employment of the various instruments of power either singly or in combination.
6. Learning Area 4—*National Security Policy Process*
   a. Comprehend the philosophical, historical and constitutional foundations of the national security establishment and process.
   b. Comprehend how domestic factors influence US national security strategy and policy.
   c. Comprehend the origins and evolving role, responsibilities, organization and *modus operandi* of the interagency process for US national security strategy and policy.
   d. Examine how US national security strategies and policies are formulated and implemented and how that process is changing over time.

    e. Examine how US resource limitations and prioritization shape national security strategies and policies.

7. Learning Area 5—*National Military Strategy*
    a. Analyze the nature of war and its evolving character and conduct—past, present and future.
    b. Apply classical and contemporary theories of war to current and future strategic challenges.
    c. Comprehend the key considerations and emerging concepts that shape the development of national military strategy.
    d. Evaluate the current national military strategy, as well as other examples of US and foreign military strategies.
    e. Comprehend the culture, organization, responsibilities and capabilities of the Military Services and the process by which operational forces are employed by combatant commanders.
    f. Comprehend the DOD process for strategic planning and assessment for both long-term and immediate security challenges.
    g. Develop an effective national military strategy for a specific security challenge and conduct strategic implementation planning.

8. Learning Area 6—*Joint Strategic Leader Development*
    a. Synthesize techniques for leading in a joint, interagency and multinational environment.
    b. Synthesize leadership skills necessary to sustain innovative, agile and ethical organizations in a joint, interagency and multinational environment.

Appendix G to Enclosure E
Industrial College of the Armed Forces (ICAF) Joint Learning
Areas and Objectives

1. Overview. ICAF studies national security strategy, with emphasis on the resource components.
2. Mission
    a. The ICAF mission is to prepare selected military and civilians for strategic leadership and success in developing our national security strategy and in evaluating, marshalling and managing resources in the execution of that strategy.
    b. ICAF contributes to the Nation's security and well being by nurturing strategic thinking and developing those critical analytical skills necessary for formulating and implementing national security decisions. The core program aims to develop senior leaders capable of critical analysis regarding national security issues and their resource component—an enhanced ability to assess a situation; ask the right questions; identify requisite reactions and consequences; and develop effective strategic solutions. The program immerses ICAF students in a joint, interagency and international environment for 10months and qualifies its graduates for JSO nomination.
3. Learning Area 1—*National Security Strategy*
    a. Evaluate how enduring philosophical and historical American principles contribute to US strategic thinking. Analyze and evaluate the foundations and operation of democratic government, the US Constitution and the design of the national security establishment.

   b. Evaluate the nature of the ever-changing domestic and international security environments and their implications for the formulation and implementation of future national security strategy.

   c. Evaluate national security organization and strategy and the instruments of national policy to achieve US objectives in peace and war to include WMD/E and terrorism by applying historical lessons learned.

   d. Evaluate alternative means for achieving national security objectives. Formulate national security strategies, with emphasis on the mobilization of national will and resources to protect and promote national interests in peace and war to include homeland defense and security.

   e. Conduct strategic assessments of selected international regions, states or issues and develop security policy options that integrate the elements of national power and the instruments of national policy in support of the national security strategy.

   f. Evaluate the capabilities and vulnerabilities of US industry and infrastructure in a global market to support national security strategy.

   g. Evaluate the impact of defense materiel acquisition policies on the US economy and the industrial base and the generation and adaptation of the military instrument of power.

   h. Evaluate the national security technological environment as an enabler for current and future competitive advantage.

4. Learning Area 2—*National Planning Systems and Processes*

   a. Evaluate the national security decision-making system and the policy formulation process and evaluate how effective they are in establishing and supporting US national security objectives.

   b. Evaluate the responsibilities and relationships of the interagency and the joint community and evaluate their implementing policies and processes for planning, organizing, coordinating and executing national security strategies.

   c. Evaluate the national economy and the national budget process.

   d. Comprehend how resource limitations and prioritization shape national security strategies and policies.

   e. Evaluate technological means, methods and processes that can lead to rapid adaptation, change and innovation in organizations to achieve competitive advantage.

5. Learning Area 3—*National Military Strategy and Organization*

   a. Synthesize national military strategies, with emphasis on mobilization and logistic requirements, across the range of military operations.

   b. Evaluate the force structure requirements and resultant capabilities and limitations of US military forces and the associated risks that affect the development of national military strategy.

   c. Apply the concepts of the strategic decision-making and defense planning processes, with emphasis on military resource requirements, in support of US national military strategy in peace and war.

   d. Evaluate the advantages derived from joint action in planning, budgeting, organizing and executing national military strategies.

   e. Evaluate the principles of joint warfare, joint military doctrine and emerging concepts to joint, interagency and multinational operations, with emphasis on the resource component in peace and war.

   f. Evaluate the resource needs, both national and international, for national defense and the processes for converting resources into US military capabilities.

6. Learning Area 4—*Theater Strategy and Campaigning*
   a. Evaluate how joint and multinational campaigns and operations support national objectives and relate to the national strategic, theater strategic and operational levels in war.
   b. Analyze the role of information operations in national security strategy and national military strategy.
   c. Synthesize joint theater strategies to meet national strategic goals, with emphasis on logistic requirements across the range of military operations.
   d. Apply an understanding of the combatant commander's perspective of the resources required to support campaign plans, to include mobilization, deployment and sustainment.
   e. Evaluate the organization, responsibilities and capabilities of military forces available to the JFCs.
   f. Apply an analytical framework that incorporates the role that factors such as geopolitics, geostrategy, society, culture and religion play in shaping the desired outcomes of policies, strategies and campaigns in the joint, interagency, and multinational arena.
7. Learning Area 5—*Joint Strategic Leader Development*
   a. Synthesize techniques for leading in a joint, interagency and multinational environment.
   b. Synthesize leadership skills necessary to sustain innovative, agile and ethical organizations in a joint, interagency and multinational environment.

Appendix H to Enclosure E
Joint and Combined Warfighting School (JCWS) Joint Professional Military Education Phase II Joint Learning Areas and Objectives

1. Overview
   a. JCWS at JFSC offers JPME Phase II for officers expected to be selected for the joint specialty. The Joint Transition Course offers a brief overview for officers entering JPME Phase II on direct entry waivers or having earned JPME Phase I equivalent credit upon graduation from an international military college.
   b. Upon arrival, JPME Phase II students should be knowledgeable of the roles and functions of their respective Service. The students should have a working knowledge of employment and sustainment requirements, including capabilities and limitations, for warfighting within their own Service. The students should also have completed a knowledge level of education in joint organizations, the Joint Strategic Planning System and the Joint Operation Planning and Execution System.
2. Mission
   a. The mission of JFSC is to: educate military officers and other national security leaders in joint, multinational and interagency operational-level planning and warfighting; and to instill a primary commitment to joint, multinational and interagency teamwork, attitudes and perspectives.
   b. JCWS instructs students on the integrated strategic deployment, employment, sustainment, conflict termination and redeployment of joint forces. The school accomplishes this through exercises and case studies in a joint seminar environment. JCWS

fosters a mutual understanding and rapport that develops when students from all Services share and challenge the ideas, values and traditions of their Services and solve joint military problems together.

   c. The goal of the Phase II program at JCWS is to build on the foundation established by the institutions teaching JPME Phase I. In addition, the faculty and student interaction in the fully joint environment of the JFSC campus cements professional joint attitudes and perspectives essential to future successful military operations.

3. Learning Area 1—*National Strategic Security Systems and Guidance, and Command Structures*

   a. Apply appropriate strategic security policies and guidance used in developing joint operational plans across the range of military operations to support national objectives.

   b. Analyze the integration of all instruments of national power in achieving strategic objectives. Focus on the proper employment of the military instrument of national power at the joint force level both as a supported instrument and as a supporting instrument of national power.

4. Learning Area 2—*Joint, Interagency and Multinational Capabilities*

   a. Synthesize the capabilities and limitations of all Services (own Service, other Services—to include SOF) in achieving the appropriate strategic objectives in joint operations.

   b. Analyze the capabilities and limitations of multinational forces in achieving the appropriate strategic objectives in coalition operations.

   c. Analyze the capabilities and limitations of the interagency processes in achieving the appropriate strategic objectives in joint operational plans.

   d. Comprehend the attributes of the future joint force and how this force will organize, plan, prepare and conduct operations.

   e. Value a thoroughly joint perspective and appreciate the increased power available to commanders through joint, combined, interagency efforts and teamwork.

5. Learning Area 3—*Information Operations*

   a. Analyze the principles, capabilities and limitations of information operations across the range of military operations and—to include pre- and post-conflict operations.

   b. Analyze the use of information operations to achieve desired effects across the spectrum of national security threats.

6. Learning Area 4—*Joint Planning*

   a. Synthesize examples of campaign/theater planning and operations. Focus on the use of planning concepts, techniques and procedures as well as integration of battlespace support systems.

   b. Analyze complex contingency operations for use of appropriate planning principles.

   c. Apply current technology, modeling, simulation and wargaming to accomplish the synchronization, employment, support and transportation planning of the joint force.

   d. Analyze the appropriate mix of battlespace support systems and functions to develop joint operational plans.

   e. Apply an analytical framework that incorporates the role that factors such as geopolitics, geostrategy, society, culture and religion play in shaping the desired outcomes of policies, strategies and campaigns in the joint, interagency, and multinational arena.

Appendix I to Enclosure E
Joint Advanced Warfighting School (JAWS) Joint Learning
Areas and Objectives

1. Overview. JAWS at JFSC focuses on the military art and science of planning, preparing and executing campaign plans for joint, interagency and multinational participants across the full range of military operations. JAWS emphasizes joint military operations at the operational and strategic level of war and crises resolution employing all instruments of national power.
2. Mission
   a. The mission of JFSC is to: educate military officers and other national security leaders in joint, interagency and multinational operational-level planning and warfighting; and to instill a primary commitment to joint, multinational and interagency teamwork, attitudes and perspectives.
   b. JAWS produces graduates who can create campaign-quality concepts, plan for the employment of all elements of national power, accelerate transformation, succeed as joint force operational/strategic planners and be creative, conceptual, adaptive and innovative. JAWS is envisioned to populate the Joint Staff and combatant commands with expertise in the joint planning processes and capable of critical analysis in the application of all aspects of national power across the full range of military operations. Students must be capable of synergistically combining existing and emerging capabilities in time, space and purpose to accomplish operational or strategic objectives.
   c. JAWS is designed for a small group of selected Service-proficient officers (O-4 to O-6) enroute to planning-related positions on the Joint Staff and in the combatant commands. Three interrelated fields of study distinguish the 10-month curriculum: Foundations in the History and Theory of War, Strategic Foundations and Operational Art and Campaigning. The school instills decision-making and complex problemsolving experience at the strategic and operational level of war with emphasis on adaptive planning processes and techniques. JAWS strives to produce "world class warfighters" by conducting graduate-level education and preparing campaign planners to operate in a chaotic environment by teaching them "how" to think.
   d. JAWS conducts single-phase education in a manner similar to other NDU Colleges and provides necessary, rigorous joint education for officers expected to be selected for the joint specialty.
   e. Services may recognize JAWS as an intermediate- or senior-level PME equivalent. JAWS meets policies applicable to intermediate- and senior-level programs.
3. Learning Area 1—*National Security Strategy, Systems, Processes and Capabilities*
   a. Analyze the strategic art to include developing, applying and coordinating the political, military, economic, social, infrastructure and informational (PMESII) elements of national power.
   b. Analyze how the constituent elements of government and American society exert influence on the national strategy process in the joint operational environment.

    c. Analyze the ends-ways-means interrelationships for achieving national security objectives.

4. Learning Area 2—*Defense Strategy, Military Strategy and the Joint Operations Concepts*

    a. Analyze the nature of war and its evolving character and conduct—past, present and future.

    b. Analyze the art and science of developing, deploying, employing and sustaining the military resources of the Nation, in concert with other instruments of national power, to attain national security objectives in a changing security environment.

    c. Evaluate the organization, responsibilities and capabilities of the Military Services (and related organizations) and the process by which operational forces and capabilities are integrated by combatant commanders.

5. Learning Area 3—*Theater Strategy and Campaigning with Joint, Interagency and Multinational Assets*

    a. Analyze joint operational art, emerging joint operational concepts and how full spectrum dominance is attained to achieve desired endstate at the least cost in lives and national treasure.

    b. Comprehend Service, joint, interagency and multinational capabilities and how these capabilities can be best integrated to attain national security objectives.

    c. Apply an analytical framework that incorporates the role that factors such as geopolitics, geostrategy, society, culture and religion play in shaping the desired outcomes of policies, strategies and campaigns in the joint, interagency, and multinational arena.

6. Learning Area 4—*Joint Planning and Execution Processes (Pre-Conflict Through Post-Conflict)*

    a. Apply contemporary and emerging planning concepts, techniques and procedures (joint operations concepts, homeland security, the effects based approach to operations, collaborative information environment, etc.) and wargaming, modeling, and simulation for integrating battlespace support systems into campaign/theater planning operations.

    b. Comprehend collaborative systems and processes employed to operationalize strategic guidance with the systematic, on-demand creation and revision of executable plans with up-to-date options in real time.

7. Learning Area 5—*Characteristics and Conduct of the Future Joint Force*

    a. Comprehend the attributes and emerging concepts of the future joint force and how this force will organize, plan, prepare and conduct operations.

    b. Analyze and evaluate techniques for leading strategic change and building consensus among key constituencies, including Service, interagency and multinational partners, given the changing nature of conflict and national security.

8. Learning Area 6—*Information Operations*

    a. Analyze the principles, capabilities and limitations of information operations across the range of military operations and plans—to include pre- and post-conflict operations.

    b. Analyze the use of information operations to achieve desired effects across the spectrum of national security threats.

9. Learning Area 7—*Joint Strategic Leader Development*

    a. Synthesize techniques for leading in a joint, interagency and multinational environment.

  b. Synthesize leadership skills necessary to sustain innovative, agile and ethical organizations in a joint, interagency and multinational environment.

Appendix J to Enclosure E
Advanced Joint Professional Military Education (AJPME) Joint Learning Areas
and Objectives

1. Overview. Advanced Joint Professional Military Education (AJPME) at JFSC is a Reserve Component (RC) course similar in content, but not identical to, the in-residence JFSC Phase II course. AJPME students shall be JPME Phase I graduates.

2. Mission. The mission of JFSC is: to educate military officers and other national security leaders in joint, multinational and interagency operational-level planning and warfighting; and to instill a primary commitment to joint, multinational and interagency teamwork, attitudes and perspectives.

   a. AJPME educates RC officers and builds upon the foundation established in JPME Phase I. It prepares RC officers (O-4 to O-6) for joint duty assignments.

   b. AJPME fulfills the requirement for RC JPME directed in DODI 1215.20, "Reserve Component Joint Officer Management Program" (reference d).

3. Learning Area 1—*National Strategic Security Systems and Guidance and Command Structures*

   a. Apply appropriate strategic security policies and guidance used in developing joint operational plans across the range of military operations to support national objectives.

   b. Analyze the integration of all instruments of national power in achieving strategic objectives. Focus on the proper employment of the military instrument of national power at the joint force level both as a supported instrument and as a supporting instrument of national power.

4. Learning Area 2—*Joint, Interagency and Multinational Capabilities*

   a. Synthesize the capabilities and limitations of all Services (own Service, other Services—to include SOF) in achieving the appropriate strategic objectives in joint operations.

   b. Analyze the capabilities and limitations of multinational forces in achieving the appropriate strategic objectives in coalition operations.

   c. Analyze the capabilities and limitations of the interagency processes in achieving the appropriate strategic objectives in joint operational plans.

   d. Comprehend the attributes of the future joint force and how this force will organize, plan, prepare and conduct operations.

   e. Value a thoroughly joint perspective and appreciate the increased power available to commanders through joint, multinational, interagency efforts and teamwork.

5. Learning Area 3—*Information Operations*

   a. Analyze the principles, capabilities and limitations of information operations across the range of military operations and plans—to include both pre- and post-conflict operations.

   b. Analyze the use of information operations to achieve desired effects across the spectrum of national security threats.

6. Learning Area 4—*Joint Planning*

a. Synthesize examples of campaign/theater planning and operations. Focus on the use of planning concepts, techniques and procedures as well as integration of battlespace support systems.
b. Analyze complex contingency operations for use of appropriate planning principles.
c. Apply current technology, modeling, simulation and wargaming to accomplish the synchronization, employment, support and transportation planning of the joint force.
d. Analyze the appropriate mix of battlespace support systems and functions to develop joint operational plans.
e. Apply an analytical framework that incorporates the role that factors such as geopolitics, geostrategy, society, culture and religion play in shaping the desired outcomes of policies, strategies and campaigns in the joint, interagency, and multinational arena.

Appendix K to Enclosure E
CAPSTONE Joint Learning Areas and Objectives

1. Overview. The CAPSTONE curriculum helps prepare newly selected G/FOs for high-level joint, interagency and multinational responsibilities. Because of its focus on joint matters and national security, as well as its completely joint student bodies and faculty, the program is thoroughly and inherently joint. The course is conducted through classroom seminars, case studies, decision exercises, local area and overseas studies and combatant command visits.
2. Mission. Ensure newly selected G/FOs understand: (1) the fundamentals of joint doctrine and the Joint Operational Art; (2) how to integrate the elements of national power in order to accomplish national security and national military strategies; and (3) how joint, interagency and multinational operations support national strategic goals and objectives.
3. Learning Area 1—*National Security Strategy*
    a. Analyze the national security policy process, to include the integration of the instruments of national power in support of the national security and national military strategies.
    b. Comprehend the impact of defense acquisition programs and policies and their implications for enhancing our joint military capabilities.
    c. Analyze the relationships between the military and cabinet-level departments, Congress, NSC, DOD agencies and the public.
4. Learning Area 2—*Joint Operational Art*
    a. Apply joint doctrine and emerging concepts.
    b. Apply joint operational art.
    c. Evaluate the processes and systems used to synchronize the effect from the application of Joint, Service, interagency, and multinational capabilities and how these capabilities can be best integrated to attain national security objectives.
    d. Analyze how Joint, Service, and multinational command and control, information operations, public affairs, and battlespace awareness are integrated to support achieving national security objectives in a Joint Operational Area (JOA).

Appendix L to Enclosure E
Combined/Joint Force Functional Component Commander
Courses Joint Learning Areas and Objectives

1. Overview. Combined/Joint Force Air Component Commander (C/JFACC), Combined/Joint Force Land Component Commander (C/JFLCC) and Combined/Joint Force Maritime Component Commander (C/JFMCC) are senior warfighting professional continuing education. These courses are owned and controlled by the Service Chiefs. The Service Chiefs delegate course development and execution to their Service executive agents: the Commandant, Army War College; President, Naval War College; President, Marine Corps University; and the Commander, Air University. Instruction for the course comes from senior national-level civilians and military representatives, flag officers serving as combatant commanders and retired, battle-tested officers. Attendees study warfighting, military doctrine and application of unified, joint and combined combat forces so they will be better prepared to face future crises as functional component commanders. Each course is approximately 1 week in length and is offered at least semi-annually. To facilitate a seminar learning experience, each class is limited to about 18 flag officers representing all Military Services.

2. Mission. The mission of the component commander courses are to prepare one, two and three-star officers of all four Services for theater-level combat leadership. They are tailored to provide future functional component commanders with a broad perspective of the operational and strategic levels of war.

3. Attendees. All attendees should be at least a one star flag officer (one star selects may attend on a case by case basis). Since these courses build on knowledge from NDU's CAPSTONE course, attendees should complete this congressionally mandated course prior to attending a component commander course. These courses are extremely high tempo, proceed incrementally and rely on the close interaction between attendees. Therefore, absences from any part of component courses are highly discouraged and need to be approved by the individual Service selection office.

4. Learning Area 1—*National Security Strategy*
   a. Analyze the relationship between political and military objectives and how the relationship may enhance or inhibit the combatant commander in reaching his theater military objectives.

5. Learning Area 2—*National Planning Systems and Processes*
   a. Comprehend joint and Service doctrine applicable to the planning and execution of operations in support of theater-level plans and operations.
   b. Comprehend how time, coordination, policy, politics, doctrine and national power affect the planning process.
   c. Apply the principal joint strategy development and operational planning processes.

6. Learning Area 3—*National Military Strategy and Organization*
   a. Comprehend the combatant commander's perspective and the role of subordinate commanders developing, deploying, employing and sustaining military forces.
   b. Comprehend the roles and functions of the component commander to include relationships with and perspectives of the Combatant Commander, Combined/Joint Force Commanders (C/JFC), and component commanders (both functional and Service).

7. Learning Area 4—*Theater Strategy and Campaigning*

   a. Comprehend the role of the unified commander in developing theater plans, policies and strategy.

   b. Comprehend the theater-level strategy development and the development of military objectives, end states and a joint concept of operations.

   c. Apply a theater component strategy that supports the C/JFC campaign plan.

   d. Comprehend the role of joint doctrine as they apply to operations planning, mobilization, deployment, employment, assessment, sustainment and redeployment.

   e. Assess issues related to component functioning (i.e., air defense, airspace coordination, theater missile defense, fire support coordination, targeting, rules of engagement, joint fires, etc.).

   f. Understand the key components, systems, and processes used to plan, direct, coordinate, control and assess combined/joint air, land, maritime and space effects-based operations.

8. Learning Area 5—*Information Operations*

   a. Comprehend how theater commanders, component commanders or JTF commanders access information operations resources and develop responsive information operations plans.

   b. Comprehend historical or on-going information operations.

   c. Comprehend the requirements necessary to collect, collate and disseminate intelligence information.

   d. Comprehend the importance of strategic communication in a multinational environment and the impact it has in shaping the information environment.

9. Learning Area 6—*The Role of Technology in 21st Century Warfare*

   a. Comprehend the role of joint experimentation, joint exercises, research and development and emerging organizational concepts with respect to transforming the US military.

   b. Comprehend the nature of warfare in the information age, to include advanced planning and analysis capabilities.

9. Learning Area 7—*Strategic Leader Development*

   a. Synthesize the unique challenges of operational command at the three-star level.

   b. Analyze the complexities associated with leadership in a coalition environment at the task force, component and combatant commander levels.

   c. Understand the complexities associated with leadership in an interagency environment at the task force, component and combatant commander levels.

Appendix M to Enclosure E
Joint Flag Officer Warfighting Course (JFOWC) Joint
Learning Areas and Objectives

1. Overview. JFOWC is an intermediate G/FO-level professional continuing education course in DOD, owned and controlled by the Service Chiefs. The Service Chiefs delegate course development and execution to their Service executive agents: the Commandant, Army War College; President, Naval War College; President, Marine Corps University; and the Commander, Air University. Instruction for the course comes from senior national-level civilians and military representatives, flag officers serving as combatant commanders and retired, battle-tested officers. Attendees study warfighting, synchronization of interagency operations, military doctrine and the application of unified,

joint and combined combat forces so they will be better prepared to face future crises. JFOWC is a 2-week course offered semi-annually. Each class is limited to 18 flag officers representing all Military Services.

2. Mission. The JFOWC mission is to prepare two-star officers of all four Services for theater-level combat leadership responsibilities. It is tailored to provide potential theater combatant commanders, Service component and JTF commanders with a broad perspective of the strategic and operational levels of war.

3. Learning Area 1—*National Security Strategy*

   a. Comprehend the role of Congress in military affairs and how Congress views the military.

   b. Comprehend the role of military leaders in developing national political objectives.

   c. Comprehend the four elements of national power: diplomatic, informational, military and economic and how the elements are used during crisis situations.

   d. Analyze the relationship between the strategic and military endstates and how they differ and influence stability operations and redeployment.

4. Learning Area 2—*National Planning Systems and Processes*

   a. Comprehend the role of joint doctrine with respect to unified command as it applies to operations planning, mobilization, deployment, employment and sustainment and redeployment.

   b. Analyze how time, coordination, policy, politics, doctrine and national power affect the planning process.

   c. Apply the principal joint strategy development and operational planning processes.

5. Learning Area 3—*National Defense Strategy*

   a. Comprehend how the military operationalize the national defense strategy to address strategic challenges by setting priorities among competing capabilities.

   b. Comprehend how the military dissuades potential adversaries from adopting threatening capabilities, methods and ambitions, particularly by sustaining and developing our own key military advantages.

6. Learning Area 4—*National Military Strategy and Organization*

   a. Comprehend the combatant commander's perspective and the role of subordinate commanders developing, deploying, employing and sustaining military forces.

   b. Analyze the roles, relationships and functions of the President, SecDef, CJCS, JCS, combatant commanders, Secretaries of the Military Departments and the Service Chiefs as related to the national military strategy.

7. Learning Area 5—*Theater Strategy and Campaigning*

   a. Examine the role of the unified commander in developing theater plans, policies and strategy.

   b. Examine the complexities of interagency coordination and support in campaign planning and execution of military operations.

   c. Examine the potential challenges and opportunities that may accrue from the combatant commander's regional focus and an ambassador's country focus.

   d. Comprehend a multinational campaign plan for a geographic combatant commander in support of national and coalition objectives.

8. Learning Area 6—*The Role of Strategic Communication in the 21st Century Warfare*

   a. Describe how theater commanders, component commanders or JTF commanders access information operations resources and develop responsive information operations plans.

    b. Comprehend the impact of national agencies that support the theater commander's requirements for information operations on national security issues.

    c. Evaluate how the joint operational planning and execution system is integrated into both theater and operational information operations campaign planning and execution to support theater and national strategic sustainment and warfighting efforts.

    d. Comprehend the importance of strategic communication in a multinational environment and the impact it has in shaping the information environment.

    e. Evaluate how public diplomacy and public affairs are integrated in theater and operational information operations planning and execution to support theater and national strategic sustainment and warfighting efforts.

9. Learning Area 7—*Strategic Leader Development*

    a. Comprehend the unique challenges of command at the three- and four-star levels.

    b. Comprehend leadership challenges in a coalition environment.

    c. Comprehend the leadership challenges in working with and understanding the cultures of other members of the interagency.

Appendix N to Enclosure E
PINNACLE Course Joint Learning Areas and Objectives

1. Overview. The PINNACLE curriculum helps prepare prospective joint/combined force commanders to lead joint and combined forces, building upon the progression of knowledge imparted first by CAPSTONE, the Combined/Joint Force Functional Component Commander Courses or the Joint Flag Officer Warfighting Course. The course is conducted through classroom interactive seminars guided by retired three- and four-star and equivalent interagency senior mentors, reinforced by video teleconferences with commanders in the field and high-level guest speakers.

2. Mission. Convey to the prospective joint/combined force commander an understanding of national policy and objectives with attendant international implications and the ability to operationalize those objectives/policy into integrated campaign plans. The overarching goal is to set conditions for future success in the joint, combined and interagency arenas by using advanced knowledge of operational art to underpin the instinct and intuition of the prospective commanders.

3. Learning Area 1—*The Joint/Combined Force Environment*

    a. Analyze the changed nature of operations for a joint/combined force commander, vis-à-vis a Service or functional component commander, identifying fundamental differences in the way a joint/combined force commander must think about the environment as its nature and complexity changes.

    b. Synthesize operational-level lessons learned from the full spectrum of recent major operations in order to evaluate them with regard to potential future operations.

    c. Evaluate the transformational concepts, including effects-based approach to operations, C2 enhancements and special operations forces integration, which will be employed in future operations.

    d. Synthesize methods to more effectively apply the diplomatic, informational, military and economic (DIME) instruments of national power to influence a potential adversary's Political, Military, Economic, Social, Infrastructure and Information (PMESII) elements.

4. Learning Area 2—*Building the Joint/Combined Force*
    a. Apply the impact of an effects-based approach across the spectrum of joint/combined force operations.
    b. Evaluate specific enablers such as effects-based planning, information/knowledge management, and battle rhythm flexibility that support the commander's decision cycle.
    c. Apply transformational concepts to traditional planning, forming, and manning options to develop alternative planning and forming options.
5. Learning Area 3—*Commanding the Joint/Combined Force*
    a. Synthesize processes to further the understanding of, identify the challenges associated with, and effectively blend the art (synergy) and science (synchronization) of commanding joint/combined forces.
    b. Apply an understanding of and appreciation for translating national objectives and policies into objectives and effects, clearly articulating the integrated tactical actions to achieve those objectives.
    c. Analyze the impact of strategic communication and information operations on unity of effort and the achievement of national objectives.
    d. Evaluate emerging technologies, which mitigate the challenges of the "death of time and distance" with regard to battlespace.
    e. Evaluate various issues related to deployment, employment and sustainment of forces from the perspective of the joint/combined force commander.
    f. Evaluate C2 challenges facing the joint/combined force commander, including the personalities of external principals (DOD, interagency, and international), transitions, and Commander's Critical Information Requirements.
    g. Analyze seams a joint/combined force commander may face, which could include interfaces with key Service, interagency, multinational and functional combatant commander representatives.
    h. Evaluate key national authority and rules of engagement issues, which could impact the joint/combined force, including national policies and prerogatives, information sharing and titles.
6. Learning Area 4—*The Joint Force Commander and the Interagency, National Command Authority, National Military Strategy and Congress*
    a. Synthesize the view of key Joint Staff members to understand the strategic view of the National Military Strategy and the required integration of the joint force commander in the interagency process with its competing interests and diverse viewpoints; both against the backdrop of current operational issues.
    b. Synthesize the view of key Department of State (DOS) officials and the role of the joint force commander in all phases of operations, specifically focusing on key DOS missions, interagency planning and the multitude of non-governmental organizations involved in execution of the national objectives and policies.
    c. Evaluate the overarching view of the intelligence services and interface with the national intelligence community leadership to understand support for fielded forces from a strategic perspective.
    d. Evaluate the role and perspective of Congress in regards to national security issues to include funding and the will of the American people as articulated by their representatives.
    e. Evaluate DOD's view of the unified commander, the joint force commander and their responsibilities to and interface with the National Command Authority.

ENCLOSURE F
PROCESS FOR ACCREDITATION OF JOINT EDUCATION (PAJE)

1. Overview. This enclosure details the charter, guidelines, preparation and conduct of the PAJE. The provisions of this enclosure apply to certification, accreditation and reaffirmation reviews. Appendix A provides the PAJE schedule, Appendix B describes the PAJE charter, and Appendix C provides guidelines for institutional self-studies required for PAJE reviews.

2. Purpose. The PAJE serves three purposes: oversight, assessment and improvement. Through the PAJE, the CJCS complies with statutory responsibilities for oversight of the officer joint educational system. The PAJE also serves as a method for improving college/schools' execution of JPME through periodic self-study and self-assessment. PAJE team assessment assures quality and assists in improvement. The PAJE is not intended to be a detailed checklist inspection of colleges/schools' programs but an opportunity for a balanced team of peers and experts to assure the Chairman that each college/school properly executes JPME and to offer the college/school the benefit of the team's findings and recommendations.

3. Background. The PAJE process is generally guided by accepted civilian accreditation standards and practices tailored to the needs of JPME. Colleges/schools teaching JPME differ from civilian universities in at least two significant ways:

   a. Underlying Theme of the Subject Matter. JPME addresses the diplomatic, informational, military and economic dimensions of national security, with special emphasis on planning and conducting activities throughout the range of military operations.

   b. Learning Environment. Colleges/schools conducting JPME bring together a faculty and student body of professional military officers and civilian government officials who have significant experience in the major disciplines taught at the colleges. Also, these colleges/schools have access to and use classified information and wargaming facilities not available to civilian universities.

4. The Process. The PAJE is a peer review process and is best accomplished by individuals with an in-depth understanding of JPME subject matter and the educational environment for ILE and SLE. Consequently, representatives (military and civilian) of the Services, Joint Staff, and NDU directly involved with JPME are selected to conduct the PAJE. Despite the PAJE team's unique composition, its concept and practice are common to all academic accreditation systems—to strengthen and sustain professional education.

5. PAJE Sequence. The sequence of PAJE reviews starts with certification, followed by accreditation, and then subsequent reaffirmation of the program's accreditation status. All PAJE reviews are conducted using the guidelines of the PAJE.

   a. Certification. Certification is the initial PAJE review and is intended for three situations: (1) programs that have never been awarded any type of PAJE accreditation status; (2) programs that were formerly certified or accredited but have had that status expire; or (3) programs that are currently certified or accredited but have undergone substantive change as defined below.

   b. Accreditation. Accreditation is the second level of PAJE review and is conducted within 2 years following an institution's certification for JPME. Accreditation is granted for 6 years when programs are judged satisfactory overall and have no significant weaknesses.

   c. Reaffirmation. Reaffirmation of accreditation occurs every 6 years from the date of initial accreditation. Reaffirmation also is granted for 6 years when programs are judged satisfactory overall and have no significant weaknesses.

   d. Conditional Accreditation/Reaffirmation. Either initial accreditation or reaffirmation can be granted on a conditional basis. Conditional accreditation or reaffirmation is granted for 1–3 years with various accompanying requirements for follow-on reports and/or follow-up visits to demonstrate correction of program weaknesses that precluded accreditation/reaffirmation. Normally, no program will be granted conditional accreditation/reaffirmation consecutively.

   e. Any program that fails to achieve accreditation, reaffirmation or conditional accreditation/reaffirmation is no longer a JPME provider.

6. Program Changes

   a. Substantive Change. The Chairman, in accordance with Paragraph 5 above, must certify again in its entirety a college or school that implements a substantive change that significantly affects the nature of the institution, its mission and objectives, its PME and/or JPME programs. Substantive change may include, but is not limited to:

      (1) Adding major PME/JPME courses or programs that depart significantly in either content or method of delivery from those offered when the college or school was most recently evaluated.

      (2) Decreasing substantially the length, hours of study, or content of a major PME/JPME course or program required for successful completion of the full course of study.

      (3) Changing the geographical setting for a resident course, to include moving to a new location, establishing a branch campus or establishing an off-campus mode of operation.

      (4) Departing significantly from the stated mission, objectives or PME/JPME programs operative at the time of the most recent evaluation.

      (5) Changing a PAJE-validated method of delivery (e.g., engaging another organization (as by contract) to provide direct instructional services).

      (6) Merging with another institution.

   b. Limited Change. A limited change to some aspect of an institution's overall program is one of sufficient extent to warrant seeking approval from the Director, Joint Staff, but not so extensive that it warrants CJCS certification of the entire program. The Director may approve a limited change based simply on the written explanation of the change or may require a validation assessment.

      (1) Validation. Validation is an interim assessment of an aspect of a college or school's program that has undergone a limited change of sufficient extent to warrant on-site evaluation but not so extensive as to warrant certification. The aim of a validation assessment is to determine whether the change falls within the scope and meets the requirements of the institution's current accreditation status or whether certification is required.

      (2) In preparing for a validation assessment, the Director may require the college or school to prepare a limited self-study or a selfstudy addendum. The on-site visit may be conducted by a member of Joint Staff/J-7 JEB or by a small team of PAJE evaluators.

   c. Advance Notification. Responsibility rests with the college or school to notify in advance the Chairman (via the chain of command) of its intent to implement a limited or substantive change and to request approval, validation or certification as appropriate. Notification should include a thorough explanation of the change's nature, extent and ramifications for the institution's PME/JPME programs. The greater the envisioned change, the further in advance notification should occur, with 12 months being the minimum notification for an envisioned substantive change. The Director, Joint Staff, may also initiate a change approval, validation or certification should he believe a college or school is implementing a limited or substantive change.

7. Scheduling of PAJE Reviews

a. Certification requests for new programs are submitted to the Chairman through the respective Service headquarters or NDU. Certification requests for formerly certified/accredited programs or substantially altered certified/accredited programs are submitted through respective channels to the DDJS-ME.

b. Requests for accreditation or reaffirmation are submitted to the DDJS-ME at least 6 months before expiration of the institution's accreditation status. Service and NDU colleges will forward their requests through their respective headquarters. Each request should indicate the specific program(s) for review and primary and alternate dates for PAJE team visits...

(1) Academic Programs. Briefly identify and describe the institution's major academic program(s).

(2) The JPME Curriculum

(a) Describe how JPME fits into the institution's academic program(s).

(b) Identify all courses that comprise the JPME curriculum. Also provide a list of guest speakers, the subject area of their presentations and how their presentations support JPME learning areas and objectives.

(c) Provide a matrix that cross-walks each JPME learning area and/or learning objective in the OPMEP to the course and lesson in the curriculum where it is addressed. (The requisite learning areas and/or learning objectives are identified in the appropriate appendix to Enclosure E of the OPMEP.)

(d) Identify any major changes planned for current course(s) and explain their effect on JPME.

(3) Curriculum Development. Describe the process used to develop and revise the JPME curriculum, to include the major participants and their roles. In particular, identify how internal and external feedback is used in revising the curriculum. Also identify the process used to ensure changes in joint doctrine and joint warfighting are incorporated into JPME.

(4) Identify noteworthy strengths or limitations concerning the institution's academic programs and curriculums.

d. Academic Evaluation and Quality Control.

(1) Explain how the college/school assesses students' success in attaining JPME objectives (see appropriate appendix to Enclosure E, OPMEP).

(2) Describe internal and external measures of assessment. Include grading procedures for students and curriculum evaluation methods for college/school effectiveness.

(3) Explain the procedures used for curriculum development for instructional standardization among seminars.

(4) List the remedial programs or assistance provided for students experiencing difficulty completing course work satisfactorily.

    (5) Describe how program curriculum deficiencies are identified and required instructional or curriculum modifications are coordinated.

    (6) Provide a copy of all instruments used to conduct follow-up surveys of students, graduates, their supervisors and the joint leadership to determine curricula and educational effectiveness of their academic programs. Identify any established procedure ensuring data obtained is used to modify the curriculum in relation to graduates' performance in the field.

    (7) Describe how the institution has acted on assessment findings in an effort to improve its effectiveness. Identify noteworthy strengths or limitations concerning the institution's academic evaluation and quality control systems.

e. Student Body

    (1) Describe the student body composition, to include affiliations by Service, department or organization; specialty code or branch (for military students); grade; average time in Service; and level of civilian and military schooling.

    (2) Identify the percentage of DOD and non-DOD civilian students within the student population.

    (3) Describe the criteria and rationale used for achieving student mixes within seminars.

    (4) Provide a breakdown of all seminars, to include student names, grade, Service, department or organizational affiliation, country and specialty code.

    (5) Identify noteworthy strengths or limitations concerning the student body.

f. Faculty

    (1) Identify JPME faculty qualifications and determine if they have appropriate credentials and experience. Identify all faculty members with any involvement with JPME, to include their function (e.g., teach, curriculum development and course director); Service, department or organizational affiliation (if appropriate); grade; area of expertise; academic degree level; military education level; and relevant joint and Service operational experience.

    (2) Describe the military faculty mix by Military Department. Include a list of all faculty designated as teaching faculty and what courses they teach.

    (3) Identify the student-to-faculty ratio for the college/school and explain how these figures were computed. Include a list of all faculty used to compute this ratio.

    (4) Describe orientation, training and updating procedures established for faculty and staff members involved in JPME development and instruction.

    (5) Describe faculty development programs available for improving instructional skills and increasing subject matter mastery in JPME (as identified in the appropriate appendix to Enclosure E, OPMEP).

    (6) Identify noteworthy strengths or limitations concerning the institution's faculty selection, qualifications, retention or development.

g. Instructional Climate

    (1) Explain how the institution ensures academic freedom, faculty and student inquiry, open exploration of ideas, lively academic debate and examination of appropriate curriculum issues.

    (2) List active and passive learning methods used by the institution and the percentage of time students are involved in each.

    (3) Describe how the institution approaches the JPME standard of joint awareness and joint perspectives. Explain what activities are used and describe how progress in this area is assessed.

(4) Identify student counseling and academic advisory services available to the students.
h. Academic Support
  (1) Library and Learning Resources Center
    (a) Describe library or learning resource operations. Include a list of library or learning resources available to students and faculty and provide examples of types of materials directly supporting JPME curriculum requirements. Comment on availability and access to joint publications, Joint Electronic Library and other resources that support JPME.
    (b) Identify noteworthy strengths or limitations in the library and its services, including: the staffing, the availability of electronic information resources, the information technology physically available, the print and non-print collections, the physical environs, the financial support's adequacy and the services provided to on-campus and off-campus students and faculty. This assessment should include results from formal and informal library surveys as well as the library administrators and staff.
  (2) Physical Resources
    (a) Describe the adequacy of the institution's physical facilities for the number of students, course offerings, faculty members and other academic requirements.
    (b) Describe the accessibility of technology and course material development resources.
    (c) Identify noteworthy strengths or limitations in physical facilities.
  (3) Financial Resources
    (a) Identify sources of financial support to the institution. Describe the adequacy of these resources to support JPME curriculum development and course execution.
    (b) Identify resource shortfalls affecting academic programs and explain how they affect the JPME curriculum.
    (c) Describe any projected changes in resource allocation affecting the JPME curriculum.

ENCLOSURE G
REFERENCES

a. Title 10, USC, section 153
b. Title 10, USC, chapter 107
c. DOD Directive 5230.9, 9 April 1996, "Clearance of DOD Information for Public Release"
d. DOD Instruction 1215.20, 12 September 2002, "Reserve Component (RC) Joint Officer Management Program"
e. DOD Instruction 1300.20, 20 December 1996, "DOD Joint Officer Management Program Procedures"
f. DOD Manual 8910.01M, June 1998, "DOD Procedures for Management of Information Requirements"
g. 2004 National Military Strategy of the United States of America. *A Strategy for Today; A Vision for Tomorrow.*

h. CJCSI 1801.01, (updated through the Joint Electronic Library), "National Defense University Education Policy", 1 July 2002
i. CJCSM 3500.04D, (updated through the Joint Electronic Library), "Universal Joint Task List (UJTL)", 1 August 2005
j. Joint Pub 1, "Joint Warfare of the Armed Forces of the United States", 10 January 1995
k. Joint Pub 1-02, (updated through the Joint Electronic Library), "DOD Dictionary of Military and Associated Terms", 31 August 2005
l. Joint Pub 3.0, (updated through the Joint Electronic Library), "Doctrine for Joint Operations"
m. Training Transformation Planning Guidance, March 2002.
n. Bloom, B. S. *Taxonomy of Educational Objectives,* 1956
o. CJCS, Capstone Concept for Joint Operations (CCJO), August 2005

GLOSSARY
PART I–ACRONYMS

ACGSS   Army Command and General Staff School

ACSC   Air Command and Staff College

AJPME   Advanced Joint Professional Military Education

AWC   Air War College

AY   Academic year

CJCS   Chairman of the Joint Chiefs of Staff

CNCS   College of Naval Command and Staff

CNW   College of Naval Warfare

DE   distance education

DDJS-ME   Deputy Director, Joint Staff, for Military Education

DJS   Director of the Joint Staff

DLCC   Distance Learning Coordination Committee

DOD   Department of Defense

GNA   Goldwater-Nichols DOD Reorganization Act of 1986

G/FO   general/flag officer

GFOCC   General and Flag Officer Coordination Committee

ICAF   Industrial College of the Armed Forces

ILC   Intermediate-level College

ILE   intermediate-level education

J-1   Directorate for Manpower and Personnel, Joint Staff

JAWS   Joint Advanced Warfighting School

JCIWS   Joint Command, Control, and Information Warfighting School

JCS   Joint Chiefs of Staff

JCSOS   Joint and Combined Staff Officer School

JCWS   Joint and Combined Warfighting School

JDA   joint duty assignment

JDAL   Joint Duty Assignment List

JEB   Joint Education Branch (JS J-7)

JFC   joint forces commander

JFEC   Joint Faculty Education Conference

JFOWC   Joint Flag Officer Warfighting Course

JFSC   Joint Forces Staff College

JLA   Joint Learning Area

JLO   Joint Learning Objective

JOPES   Joint Operation Planning and Execution System

JOM   Joint Officer Management

JPME   Joint professional military education

JS   Joint Staff

JS   J-7 Directorate for Operational Plans and Joint Force Development, Joint Staff

JSPS   Joint Strategic Planning System

JSO   joint specialty officer

JTF   joint task force

LA   learning areas

LO   learning objectives

MCCCE   Marine Corps College of Continuing Education

MCCSC   Marine Corps Command and Staff College

MCWAR   Marine Corps War College

MECC   Military Education Coordination Council

NDU   National Defense University

NPS   Naval Postgraduate School

NSC   National Security Council

NWC   National War College

OCS   officer candidate school

OPMEP   Officer Professional Military Education Policy

OSD   Office of the Secretary of Defense

OTS   officer training school

PAJE   Process for Accreditation of Joint Education

PME   professional military education

POI   Program of Instruction

POM   program objective memorandum

RC JPME   Reserve Component Joint Professional Military Education

ROTC   Reserve Officer Training Corps

SAE   special area of emphasis

SIWS   School of Information Warfare and Strategy

SLC   Senior-Level College

SLE   senior-level education

UJTL   Universal Joint Task List

USAWC   US Army War College

VDJ-7   Vice Director for Operational Plans and Joint Force Development

WMD   Weapons of Mass Destruction

WMD/E   Weapons of Mass Destruction/Effects

PART II–DEFINITIONS

academic freedom—Freedom to pursue and teach relevant knowledge and to discuss it freely as a citizen without interference, as from school or public officials.

accreditation—The granting of approval to an institution of learning by the CJCS after the school has satisfied the requirements specified in the Process of Accreditation for Joint Education (PAJE). Accreditation is the second level of PAJE review and is conducted within 2 years following an institution's certification for JPME. Accreditation is granted for 6 years when programs are judged satisfactory overall and have no significant weaknesses.

CAPSTONE—CAPSTONE is a mandated 6-week course for newly selected G/FOs. The course objective is to make these individuals more effective in planning and employing US forces in joint and combined operations. The CAPSTONE curriculum examines major issues affecting national security decision-making, military strategy, joint and combined doctrine, interoperability and key-allied nation issues.

certification—Certification is the assessment of a college or school as to whether it meets JPME requirements. Certification provisionally accredits a program for 2 years or until a full accreditation occurs. Certification is used in three situations: (1) programs that have never been awarded any type of PAJE accreditation status; (2) programs that were formerly certified or accredited but have had that status expire; or (3) programs that are currently certified or accredited but have undergone substantive change.

conditional accreditation/reaffirmation—Initial accreditation or reaffirmation can be granted on a conditional basis. Conditional accreditation or reaffirmation is granted for 1 to 3 years with various accompanying requirements for follow-on reports and/or follow-up visits to demonstrate correction of program weaknesses that precluded accreditation/reaffirmation. Normally, no program will be granted conditional accreditation/reaffirmation consecutively.

culture—The distinctive and deeply rooted beliefs, values, ideology, historic traditions, social forms, and behavioral patterns of a group, organization, or society that evolves, is learned, and transmitted to succeeding generations.

cultural knowledge—Understanding the distinctive and deeply rooted beliefs, values, ideology, historic traditions, social forms, and behavioral patterns of a group, organization, or society; understanding key cultural differences and their implications for interacting with people from a culture; and understanding those objective conditions that may, over time, cause a culture to evolve.

direct-entry waiver—A waiver, requested by a Service and approved by the Chairman of the Joint Chiefs of Staff, that permits an officer who is neither a graduate from a certified or accredited JPME Phase I course of instruction nor a recognized Phase I-equivalent program, to attend JPME Phase II prior to completion of Phase I. The waiver only concerns the sequencing of the JPME phases and does not alter the requirement for completion of both JPME phases to meet the full education prerequisite for JSO/JSO nominee designation. *(DODI 1300.20)*

distance education—learning situation in which the instructor and/or students are separated by time, location, or both. Education or training courses are delivered to remote locations via synchronous or asynchronous means of instruction, including written correspondence, text, graphics, audio- and videotape, CD-ROM, distributed online learning, audio- and videoconferencing and fax. Distance education does not preclude the use of the traditional classroom. The definition of "distance education" is usually meant to describe something, which is broader than and entails the definition of e-learning.

Distance Learning Coordination Committee (DLCC)—primary advisory body to the MECC WG on DL issues. The DLCC is an ongoing forum to promote best practices, exchange shareware, and provide and exchange information regarding technical and nontechnical issues in Distance Learning in order to encourage collaboration, joint enterprise, and leverage of membership successes. Membership consists of the Deans and Directors of all Distance Education programs at the intermediate and senior-level PME institutions with distance learning programs, encompassing continuing education and nonresident PME programs at the various Service and Joint education institutions. Assignment or appointment in the positions confirms membership. Other military education institutions, not members of the MECC, may apply for associate membership and participate in DLCC activities. The DLCC briefs the MECC on issues of concern as appropriate.

distributed—refers to the capability for institutions to use common standards (OSD Advanced Distributed Learning initiative, for example, Shareable Content Object Reference Model (SCORM)) and network technologies in order to provide learning anywhere and anytime.

education—Education conveys general bodies of knowledge and develops habits of mind applicable to a broad spectrum of endeavors.

e-learning—Broad definition of the field of using technology to deliver education and training programs. It is typically used to describe media such as DVD, CD-ROM, Internet, Intranet, or wireless learning.

faculty—Personnel (military or civilian) who teach, conduct research or prepare or design curricula.

General and Flag Officer Coordination Committee (GFOCC)—primary advisory body to the MECC WG on G/FO issues. This MECC WG subgroup was created to: integrate the individual efforts regarding the education of G/FOs; discuss common areas of interest, establish a community of interest, G/FO education network and to chart a vision for the future.

individual joint training—Training that prepares individuals to perform duties in joint organizations (e.g., specific staff positions or functions) or to operate uniquely joint systems (e.g., joint intelligence support system). Individual joint training can be conducted by OSD, the Joint Staff, combatant commands, Services, Reserve Forces, National Guard or combat support agencies. *(Joint Training Policy)*

intermediate-level education (ILE)—A formal, intermediate-level Service college; includes institutions commonly referred to as intermediate Service colleges, intermediate-level schools, intermediate Service schools or military education level-4 producers.

Joint Advanced Warfighting School (JAWS)—Course designed to produce graduates that can create campaign-quality concepts, employ all elements of national power, accelerate transformation, succeed as joint force operational/strategic planners and commanders and be creative, conceptual, adaptive and innovative. JAWS is envisioned to populate the Joint Staff and combatant commands with officers expert in the joint planning processes and capable of critical analysis in the application of all aspects of national power across the full range of military operations.

Joint Flag Officers Warfighting Course (JFOWC)—JFOWC prepares two-star officers of all four services for theater-level combat leadership responsibilities. It is tailored to provide future theater combatant commanders, Service component and JTF commanders with a broad perspective of the strategic and operational levels of war.

Joint Professional Military Education (JPME)—A CJCS-approved body of objectives, outcomes, policies, procedures and standards supporting the educational requirements for joint officer management.

JPME phases—A three-phase joint education program taught at Service intermediate- or senior-level colleges, Joint Forces Staff College and NDU for the CAPSTONE course that meets the educational requirements for joint officer management.

(a) JPME Phase I—A first phase of JPME is incorporated into the curricula of intermediate- and senior-level Service colleges and other appropriate educational programs, which meet JPME criteria and are accredited by the CJCS. By law, the subject matter to be covered shall include at least the following: (1) national military strategy; (2) joint planning at all levels of war; (3) joint doctrine; (4) joint command and control, and (5) joint force and joint requirements development.

(b) JPME Phase II—A follow-on second phase of JPME for selected graduates of Service schools and other appropriate education programs that complements and enhances Phase I instruction. This phase is taught at JFSC JCWS to both intermediate- and senior-level students, at Service senior-level colleges to senior-level students and completes their educational requirement for joint officer management. In addition to the subjects specified in JPME Phase I above, by law, the curriculum for Phase II JPME shall include the following: (1) National security strategy; (2) theater strategy and campaigning; (3) joint planning processes and systems, and (4) joint, interagency and multinational capabilities and the integration of those capabilities.

(c) CAPSTONE—by law, the third phase of JPME. joint training—Military training based on joint doctrine or JTTP to prepare individuals, joint commanders, joint staff and joint forces to respond to strategic and operational requirements deemed necessary by combatant commanders to execute their assigned missions. Joint training involves forces of two or more Military Departments interacting with a combatant commander or subordinate joint force commander; involves joint forces and/or joint staffs; and or individuals preparing to serve on a joint staff or in a joint organization and is conducted using joint doctrine or TTP. *(Joint Training Policy)*

Military Education Coordination Council (MECC)—An advisory body to the Director, Joint Staff, on joint education issues, consisting of the MECC Principals and a supporting MECC Working Group. The purpose of the MECC is to address joint scholarship and key educational issues of interest to the joint education community, promote cooperation and collaboration among the MECC member institutions and coordinate joint education initiatives.

Military Education Coordination Council Principals—The MECC Principals are the: DDJS-ME; the Presidents, Directors and Commandants of the JPME colleges, Service universities, ILCs, and SLCs; USJFCOM/J-7 and the heads of any other JPME certified or accredited institutions.

Military Education Coordination Council Working Group—A working group comprised of representatives (O-6s and dean-level civilian counterparts) of the MECC Principals. Chief, Joint Staff/J-7 JEB, chairs the working group. Its primary function is coordination of MECC agenda items.

nonresident education—The delivery of a structured curriculum to a student available at a different time or place than the teaching institution's resident program. There are three approaches used to provide nonresident JPME via an appropriate, structured curriculum: satellite seminars or classes, distance/distributed learning and blended learning.

PINNACLE—Course designed to prepare senior G/FOs for senior political-military positions and command of joint and coalition forces at the highest level. It is designed to sensitize them to the environment in which they are about to enter as well as foster understanding of national and international objectives, policies and guidance.

Process for Accreditation of Joint Education (PAJE)—A CJCS-approved process for oversight, assessment, and improvement of the JPME programs at intermediate and senior colleges.

professional military education (PME)—PME conveys the broad body of knowledge and develops the habits of mind essential to the military professional's expertise in the art and science of war.

range of military operations (ROMO)—A doctrinal term (Joint Pub 3–0), and a conceptual term used in joint concepts (such as the Capstone Concept for Joint Operations—CCJO), it consists of broad categories—and types of operations—Military Engagement/Security Cooperation & Deterrence, Crisis Response contingencies, and Major Operations & Campaigns (both adversary-focused and humanitarian/nonadversary ops). See Joint Pub 3–0 or CCJO for examples of the activities and specific operations that may be included under this definition.

reaffirmation—A follow-on accreditation review of an institution to determine whether it continues to meet PAJE standards. Reaffirmation of accreditation occurs every 6 years from the date of initial accreditation. Reaffirmation also is granted for up to 6 years when programs are judged satisfactory overall and have no significant weaknesses.

senior-level education (SLE)—A formal, senior-level Service or NDU college; includes institutions commonly referred to as top-level schools, senior Service colleges, senior Service schools or military education level-1 producers.

Shareable Content Object Reference Model (SCORM)—a collection of specifications that defines a web-based learning "Content Aggregation Model," "Run-time Environment" and "Sequencing and Navigation" for reusable content objects. At its simplest, it is a model that references a set of interrelated technical specifications and guidelines designed to meet the DOD's high-level requirements for e-learning content.

single-phase JPME—curricula reflecting the distinct educational focus and joint character of NWC, ICAF and JAWS missions.

weapons of mass destruction/effects. WMD/E—relates to a broad range of adversary capabilities that pose potentially devastating impacts. WMD/E includes chemical, biological, radiological, nuclear and enhanced high explosive weapons as well as other, more asymmetrical "weapons." They may rely more on disruptive impact than destructive kinetic effects. For example, cyber attacks on US commercial information systems or attacks against transportation networks may have a greater economic or psychological effect than a relatively small release of a lethal agent. They also include threats in cyberspace aimed at networks and data critical to US information-enabled systems. Such threats require a comprehensive concept of deterrence encompassing traditional adversaries, terrorist networks and rogue states able to employ a range of offensive capabilities.

Quadrennial Defense Review Report

A document which offers guidance and vision for the Defense Department's goals with regard to professional military education is the Quadrennial Defense Review Report. While excerpted, the key goal of jointness and more coordinated activities is absolutely clear in this document which Congress mandates must appear every four years.

Quadrennial Defense Review (Washington, D.C.: U. S. Government Printing Office, 2006) excerpted

No prudent military commander wants a fair fight, seeking instead to "overmatch" adversaries in cunning, capability and commitment. The selfless service and heroism of the men and women of the well-trained all-volunteer Total Force has been a primary source of U.S. strategic overmatch in confronting the wide range of threats we face and a key to successful military operations over the past several decades. The Total Force must continue to adapt to different operating environments, develop new skills and rebalance its capabilities and people if it is to remain prepared for the new challenges of an uncertain future. Recent operational experiences highlight capabilities and capacities that the Department must instill in the Total Force to prevail in a long, irregular war while deterring a broad array of challenges. The future force must be more finely tailored, more accessible to the joint commander and better configured to operate with other agencies and international partners in complex operations. It must have far greater endurance. It must be trained, ready to operate and able to make decisions in traditionally nonmilitary areas, such as disaster response and stabilization. Increasing the adaptability of the Total Force while also reducing stress on military personnel and their families is a top priority for the Department. These imperatives require a new strategy for shaping the Department's Total Force, one that will adjust policies and authorities while introducing education and training initiatives to equip civilian and military warfighters to overmatch any future opponent.

The Department and Military Services must carefully distribute skills among the four elements of the Total Force (Active Component, Reserve Component, civilians and contractors) to optimize their contributions across the range of military operations, from peace to war. In a reconfigured Total Force, a new balance of skills must be coupled with greater accessibility to people so that the right forces are available at the right time. Both uniformed and civilian personnel must be readily available to joint T is operational Total Force must remain prepared for complex operations at home or abroad, including working with other U.S. agencies, allies, partners and nongovernmental organizations. Routine integration with foreign and domestic counterparts requires new forms of advanced joint training and education. Finally, the Department must effectively compete with the civilian sector for high-quality personnel. The transformation of the Total Force will require updated, appropriate authorities and tools from Congress to shape it and improve its sustainability. Two key enablers of this transformation will be a new *Human Capital Strategy* for the Department, and the application of the new National Security Personnel System to manage the Department's civilian personnel.

Reconfiguring the Total Force

Recent operational experiences in Iraq and Afghanistan highlight the need to rebalance military skills between and within the Active and Reserve Components. Accordingly, over the past several years, the Military Departments are rebalancing—shifting, transferring or eliminating—approximately 70,000 positions within or between the Active and Reserve Components. The Department plans to rebalance an additional 55,000 military personnel by 2010.

The Military Departments are applying this same scrutiny across the Total Force to ensure that the right skills reside inside each element. The Military Departments and Combatant Commanders will continually assess the force to ensure it remains responsive to meet future demands. U.S. Joint Forces Command (U.S. JFCOM), as the joint force provider, is aiding the effort by ensuring the appropriate global distribution of ready forces and competencies. The Department plans to introduce a new methodology and review process to establish a baseline for personnel policy, including the development of joint metrics and a common lexicon to link the Defense Strategy to Service-level rebalancing decisions. This process will help synchronize rebalancing efforts across the Department.

A Continuum of Service

The traditional, visible distinction between war and peace is less clear at the start of the 21st century. In a long war, the United States expects to face large and small contingencies at unpredictable intervals. To fight the long war and conduct other future contingency operations, joint force commanders need to have more immediate access to the Total Force. In particular, the Reserve Component must be operationalized, so that select Reservists and units are more accessible and more readily deployable than today. During the Cold War, the Reserve Component was used, appropriately, as a "strategic reserve," to provide support to Active Component forces during major combat operations. In today's global context, this concept is less relevant. As a result, the Department will: Pursue authorities for increased access to the Reserve Component: to increase the period authorized for Presidential Reserve Call-up from 270 to 365 days.

Better focus the use of the Reserve Components' competencies for homeland defense and civil support operations, and seek changes to authorities to improve access to Guard and reserve consequence management capabilities and capacity in support of civil authorities. Achieve revision of Presidential Reserve Call-Up authorities to allow activation of Military Department Reserve Components for natural disasters in order to smooth the process for meeting specific needs without relying solely on volunteers. Allow individuals who volunteer for activation on short notice to serve for long periods on major headquarters staffs as individual augmentees. Develop select reserve units that train more intensively and require shorter notice for deployment. Additionally, the Military Departments will explore the creation of all-volunteer reserve units with high-demand capabilities, and the Military Departments and Combatant Commanders will expand the concept of contracted volunteers.

Building the Right Skills

Maintaining the capabilities required to conduct effective multi-dimensional joint operations is fundamental to the U.S. military's ability to overmatch adversaries. Both battlefield integration with interagency partners and combined operations—the integration of the joint force and coalition forces—will be standard features in future operations. The combination of joint, combined and interagency capabilities in modern warfare represents the next step in the evolution of joint warfighting and places new demands on the Department's training and education processes.

Joint Training

The QDR assessed and compared the joint training capabilities of each of the Military Departments. Although the Military Departments have established operationally proven processes and standards, it is clear that further advances in joint training and education are urgently needed to prepare for complex, multinational and interagency operations in the future. Toward this end, the Department will:

Develop a Joint Training Strategy to address new mission areas, gaps and continuous training transformation.

Revise its Training Transformation Plan to incorporate irregular warfare, complex stabilization operations, combating WMD and information operations.

Expand the Training Transformation Business Model to consolidate joint training, prioritize new and emerging missions and exploit virtual and constructive technologies.

Language and Cultural Skills

Developing broader linguistic capability and cultural understanding is also critical to prevail in the long war and to meet 21st century challenges. The Department must dramatically increase the number of personnel proficient in key languages such as Arabic, Farsi and Chinese and make these languages available at all levels of action and decision—from the strategic to the tactical. The Department must foster a level of understanding and cultural intelligence about the Middle East and Asia comparable to that developed about the Soviet Union during the Cold War. Current and emerging challenges highlight the increasing importance of Foreign Area Officers, who provide Combatant Commanders with political military analysis, critical language skills and cultural adeptness. The Military Departments will increase the number of commissioned and non-commissioned officers seconded to foreign military services, in part by expanding their Foreign Area Officer programs. This action will foster professional relationships with foreign militaries, develop in-depth regional expertise, and increase unity of effort among the United States, its allies and partners. Foreign Area Officers will also be aligned with lower echelons of command to apply their knowledge at the tactical level. To further these language and culture goals, the Department will:

Increase funding for the Army's pilot linguist program to recruit and train native and heritage speakers to serve as translators in the Active and Reserve Components.... Require language training for Service Academy and Reserve Officer Training Corps scholarship students and expand immersion programs, semester abroad study opportunities and inter-academy foreign exchanges.

Increase military special pay for foreign language proficiency. Increase National Security Education Program (NSEP) grants to American elementary, secondary and post-secondary education programs to expand non-European language instruction. Establish a Civilian Linguist Reserve Corps, composed of approximately 1,000 people, as an on-call cadre of high-proficiency, civilian language professionals to support the Department's evolving operational needs.

Modify tactical and operational plans to improve language and regional training prior to deployments and develop country and language familiarization packages and operationally-focused language instruction modules for deploying forces.

Training and Educating Personnel to Strengthen Interagency Operations The ability to integrate the Total Force with personnel from other Federal Agencies will be important to reach many U.S. objectives. Accordingly, the Department supports the creation of a National Security Officer (NSO) corps—an interagency cadre of senior military and civilian professionals able to effectively integrate and orchestrate the contributions of individual government agencies on behalf of larger national security interests. Much as the Goldwater-Nichols requirement that senior officers complete a joint duty assignment has contributed to integrating the different cultures of the Military Departments into a more effective joint force, the QDR recommends creating incentives for senior Department and non-Department personnel to develop skills suited to the integrated interagency environment. The Department will also transform the National Defense University, the Department's premier educational institution, into a true National Security University. Acknowledging the complexity of the 21st century security environment, this new institution will be tailored to support the educational needs of the broader U.S. national security profession. Participation from interagency partners will be increased and the curriculum will be reshaped in ways that are consistent with a unified U.S. Government approach to national security missions, and greater interagency participation will be Bringing all the elements of U.S. power to bear to win the long war requires overhauling traditional foreign assistance and export control activities and laws. These include foreign aid, humanitarian assistance, post-conflict stabilization and reconstruction, foreign police training, International Military Education and Training (IMET) and, where necessary, providing advanced military technologies to foreign allies and partners. In particular, winning the long war ... requires strengthening the Department's ability to train and educate current and future foreign military leaders at institutions in the United States. Doing so is critical to strengthening partnerships and building personal relationships. In all cases, they are integral to successful irregular warfare operations. For example, quick action to relieve civilian suffering, train security forces to maintain civil order and restore critical civilian infrastructure denies the enemy opportunities to capitalize on the disorder immediately following military operations and sets more favorable conditions for longer term stabilization, transition and reconstruction. Full integration of allied and coalition capabilities ensures unity of effort for rapidly evolving counterinsurgency operations. Similarly, foreign leaders who receive U.S. education and training help their governments understand U.S. values and interests, fostering willingness to unite in a common cause. . . . Expand the Counter Terrorism Fellowship Program beyond its current focus on senior-level government officials and national strategic issues. Combatant Commanders and U.S. Chiefs of Mission, in consultation with regional partners, will develop education programs to improve regional counter-terrorism campaigns and crisis response planning.

Glossary

1903 General Staff Act A law authorizing the creation of the Army War College and to pay for the construction of Roosevelt Hall at the Washington Barracks. This created a broad shift toward professional military education across the services.

1947 National Security Act A major reform for the national security community after World War II, creating the Department of Defense out of the Department of War and the Joint Chiefs of Staff to represent the services as well as commands. It also created the Central Intelligence Agency, the National Security Council and National Security Advisor.

1986 Goldwater-Nichols Military Reform Act The last major defense reform that had a profound effect on the concept of a "joint" military instead of competitive, redundant service capacities. It also directly created a new sense of joint professional military education.

1990 Defense Acquisition Workforce Improvement Act The law which took the study and personnel management of the acquisition function, a major portion of the Defense Department's mobilization and logistics concerns, into the classrooms of the Industrial College of the Armed Forces.

1992 National Defense Authorization Act The Congress dictated that all officers gaining their commissions after September 30, 1996, would serve under reserve commissions, rather than regular commissions.

1996 National Defense Authorization Act Congress established the period of obligation for students who go through their undergraduate educational experiences by receiving federal tax dollars as a six-year commitment.

AFIADL *see* Air Force Institute for Advanced Distributed Learning.

Air Force Institute for Advanced Distributed Learning The Air Force institution with the greatest, most advanced commitment to PME in a distance-learning setting.

Air Force Virtual Education Center   The distance education facility allowing the Air Force to have a robust remote-learning experience.

All Volunteer Force   With President Nixon's decision to terminate the lottery draft system in 1973, the U.S. military became an all-voluntary system known as the All Volunteer Force by which officers and enlisted personnel self-select as participants in national military service.

AVF   *see* All Volunteer Force.

AVFEC   *see* Air Force Virtual Education Center.

Bonuses   At various points in the U.S. military experience, too few recruits were brought into the services (especially the Army) so a bonus program began as an incentive to bring more people into national service, often including educational benefits.

Cadet Honor Code   The standards of behavior established by the service academies for their students, with the inherent understanding that service may require meeting a standard of honor above and beyond that of civilian society.

Cadet Leadership Development System   Created in the mid-1990s to replace the hazing system for first-year students that had been in place at West Point for a century, this leadership development was established to create a bond and a cadre of similar experiences in leadership for each class at the Academy.

Chain of command   The armed forces operate with a hierarchical system of ranks that offer a clear-cut, elaborated ladder of responsibility based on rank from general or flag officer to enlisted personnel. It is therefore a chain of command operation.

Chief of Naval Operations   The senior officer in the Navy chain of command is the chief of naval operations. This individual holds a four-star rank and is responsible for setting the policy and operations for the Navy. As the service has become more complicated over the years, the freedom and personal choices of the chief have become somewhat limited, as true of the other services as well.

Chiefs of Staff   The senior person in each of the services with the responsibility for setting policy and being the most prominent "face" of the service is the chief of staff. The Marine Corps refers to its chief of staff as the commandant, while the Air Force and Army retain the titles of chief of staff. The senior figure in the Navy is the chief of naval operations. These individuals, all holding four-star rank, receive two-year (renewable) appointments by the president of the United States that are confirmed by the Senate.

CINCs   The use of this term ended under the administration of President George W. Bush when Secretary of Defense Donald Rumsfeld noted that the nation has a single commander in chief. CINCs (pronounced sinks) is the slang military personnel used for the commanders in chief (or more often commander-in-chiefs {sic}) that are the senior officers in a combatant command, these being functional like joint force command or regional such as the southern command. Congress confers the four-star rank on these senior officers upon the individual's nomination by the president for a three-year stint as the commander. The CINCs or commanders in chief together constitute the Joint Chiefs of Staff.

The Citadel   Formally known as the Military College of South Carolina, the Citadel, located in Charleston, has operated since the mid-nineteenth century, providing a significant number of the newly commissioned officers of the U.S. Army each year. Not formally part of the U.S. entry-level educational institutions, the Citadel operates much as the U.S. Military Academy at West Point, New York.

Citizen-soldier   A fairly isolated view to the United States, this eighteenth-century concept argues for a nonprofessional military officer but one who is both a citizen and a solider, educated but efficient in his/her duties without making military life the sole work one accomplishes. Upon completing a mission, citizen Soldiers return to their regular careers.

Clements Committee on Excellence in Education   Headed up by Deputy Secretary of Defense William Clements in the mid-1970s, this group recommended consolidation of curricula at the National War College and Industrial College of the Armed Forces that resulted in the National Defense University in 1976 but has never created the unified curricula that Clements suggested.

Commander in Chief   The military term for the president of the United States, the emphasis is on the Constitutional directive that he/she commands the entire military of the nation. In common usage until the administration of President George W. Bush, it was also the term for combatant commanders at the regional commands. Used only for the President after a declaration by Secretary of Defense Donald Rumsfeld, its common usage pronounces it "sink." See also CINC.

CNO   *See* Chief of Naval operations.

Distance Learning   Nonresident educational courses available online.

Dodge Commission   After some concerns resulting from U.S. performance in the Spanish-American War, Major General Grenville Dodge headed a commission to evaluate the quality performance of the War Department. Part of the result of its findings was President Theodore Roosevelt's selection of Elihu Root as secretary of war, leading to major reforms.

Draft   The system by which men in the United States were selected to serve in uniform. The draft was conducted in a lottery format, based on birthdates, for much of the mid-twentieth century to ensure a somewhat equal chance of being asked to serve the nation.

Education versus training   Professional military education is one of the processes through which the armed services offer their officers the analytical skills that can be used in various circumstances and settings, while training is the acquired skill for a specific application such as on a plane or using a particular weapons system.

Foreign Area Officer Program (FAOs)   The Foreign Area Officer Program, most well developed in the Army system, allows an officer to develop area specific expertise along the lines of the traditional system of area studies in which the officer develops a language expertise along with economic, military, geographic, political, cultural, and social understanding to operate as a military attache in that particular region. The Navy, Marine Corps, and Air Force may also create serious FAO programs as a result of the 2006 Quadrennial Defense Review recommendations but do not hold them at present.

Fourth Class Plebe System   This was an institutionalized hazing of the youngest class at the U.S. Military Academy in West Point. Intended to instill discipline and respect in the junior most members of the Corps of Cadets, the system was controversial because it appeared a form of ritualized abuse to many outsiders. The superintendant of West Point in 1990, General David Palmer, ended the practice. His successor, General Howard Graves, replaced the system with the Cadet Leader Development System.

HASC   The House of Representatives Armed Service Committee, known as the "hasque," is the focus of military issues in the lower body of the Congress. It was instrumental to the reform of the military and its educational system.

Holloway Plan   Begun in the immediate aftermath of World War II, this program began the Navy's involvement in the Reserve Office Training Program, known as NROTC.

ICAF   Industrial College of the Armed Forces that is a single-phase JPME 10-month master's degree program, concentrating on logistics and mobilization.

ILCs   Intermediate-level colleges

JAWS   Joint Advanced Warfighting School that is a single-phase JPME program.

Joint Doctrine   Fundamental principles that guide the employment of forces of two or more services in coordinated action toward a common objective. It will be promulgated by the chairman of the Joint Chiefs of Staff, in coordination with the combatant commands, services, and Joint Staff.

Joint Duty Assignment List   The list of approved positions that fulfill the 1986 Law requirement that jointness be promoted for officers who have attended Joint Professional Military Education and can only be met by officers with the aforementioned joint education.

Joint Professional Military Education   Approved curricula at certified institutions within the U.S. military education system that will contribute to the creation of joint specialty officers.

Joint Specialty Officers   Established by Title IV of the Goldwater-Nichols Defense Reform of 1986, these are officers that have a general orientation beyond their individual Service to support greater joint perspectives and issues.

Joint Tactics, Techniques, and Procedures   The actions and methods that implement joint doctrine and describe how forces will be employed in joint operations. They will be promulgated by the chairman of the Joint Chiefs of Staff, in coordination with the combatant commands, services, and Joint Staff.

JPME   *see* Joint Professional Military Education.

JSOs   *see* Joint Specialty Officers.

METL   Mission-essential Task List includes the items the unit must be able to accomplish, which are thus linked to the education and training of the military unit.

Military Education Coordinating Committee   A group representing all of the institutions with military education within the Defense Department, which meets quarterly and sets policy by its coordinating activities.

Military Officers' Education Program   *see* Reserve Officer Training Corps.

NWC   Either the National War College in Washington, D.C., which is a ten-month master's degree course in national security strategy, or the Naval War College in Newport, Rhode Island, which is a master's degree in naval science over a ten-month program.

0–5   Officer at fifth rank, or lieutanant colonel in Army, Marine Corps and Air Force while a commander in the Navy. Ranks for officers go from 0–1 to 0–10.

Officer Candidate School (OCS)   This is a Navy accession program for officers who have proven themselves as enlisted sailors or non-commissioned officers but have gotten regular commissions and have not had the opportunity to study at the U.S. Naval Academy in Annapolis, Maryland. OCS is also a Marine Corps program for college graduates who had not attended a service academy or were noncommissioned officers who had decided to take the Marine officer career path.

Officer Training School (OTS)   An Air Force program highlighting issues that a commissioned officer through the Air Force Academy in Colorado Springs would cover, for those who are entering their commissions through a non-academy path.

OPMEP   This document, issued by the chairman of the Joint Chiefs of Staff, is the most authoritative guidance on officer professional military education. Its name, spelled out, is Officer Professional Military Education Policy. The most recent one was issued on December 22, 2005.

Packard Commission on Military Reform   The 1985 blue ribbon committee, headed by industrialist David Packard, looked at management of the Defense Department under President Ronald Reagan.

PAJE   Process for Accreditation of Joint Education, carried out at PME institutions by committees of specialists set forth by the J-7 of the Joint Staff.

POTUS   *Slang* for the president of the United States.

RC   *see* Reserve Component.

Reserve Component   The Reserves in the U.S. military are collectively called the Reserve Component.

Reserve Officer Training Corps (ROTC)   This program began in 1916 when the Congress formalized it as an accession path for officers entering the Force in all Services. It had its antecedents in the Morrill Land Grant Act of 1862, which had land-grant universities include engineering, military tactics, and agriculture in their curricula.

Richardson Committee   A group, under Admiral James O. Richardson, asked to evaluate proposals for joint professional military education in the aftermath of World War II.

Root Reforms   Changes to the armed forces, particularly the Army, instituted by Secretary of War Elihu Root during his tenure in the Roosevelt administration. These led to considerably greater emphasis on education and an army that could be used in the newly acquired territories of the Pacific and Caribbean.

ROTC   *see* Reserve Officer Training Corps.

SASC   *Slang* for the Senate Armed Services Committee, pronounced "Sass K"

SecDef   *Slang* for the secretary of defense.

Senate Armed Services Committee   The Senate committee with oversight in military issues, allowing for some overlap with other standing Senate committees such as the Veterans' Affairs.

Service Secretaries   Civilian, political appointees who nominally head the Air Force, Army, and Navy. Before the 1947 reforms, they were more powerful than they are currently.

Skelton Committee Reforms   These were changes in the professional military education system recommended by Missouri Democratic Congressman Ike Skelton, a World War II veteran, with a strong interest in professional military education in the United States. Important in late 1980s and 1990s.

SLCs   Senior-level colleges, having a Service basis rather than a joint basis.

Standing army   Resulting from the philosophical debate in the late colonial and early independence period of the United States, the country eventually moved to the idea of a professional army with the need for specialized career paths and educational opportunities to enhance the force's ability to analyze its requirements to better achieve national security goals.

"Stop loss" policy   Instituted at times of great stress for the services, "stop loss" has been invoked during the Iraq war. The policy prevents a service member from retiring upon completion of his/her tour of duty. The Army, in particular, has invoked "stop loss" to retain necessary forces for the Iraq and Afghan actions.

Total Force   Under the volunteer military begun in 1973, this concept envisions the use of the entire military, integrated into a single, coherent fighting machine. Emphasizes active duty, National Guard and Reserves together.

Unity of command   A single commander calls the shots for all applicable units to achieve their unified goal.

Universal Military Training   Mandatory military education and thus service for the nation.

Virginia Military Institute   Dating back to the early years of the nineteenth century in Lexington, Virginia, the Virginia Military Institute has graduated a significant number of men and women who applied for military commissions to serve the nation. General George Catlett Marshall earned his degree at VMI before ultimately becoming Army chief of staff during the Franklin Delano Roosevelt administration during World War II.

War gaming   A popular form of education in the Services where a scenario is played by students with clear-cut educational goals in mind.

Resources

The resources in this chapter are not exhaustive but attempt to illustrate the range of sources for military education information. The topic tries to remain faithful to the concept of professional military education (PME) although the public and many decision makers use the terms "education" and "training" interchangeably. Professional educators dispute this concept that the two are the same but some of the resources here include training in the title because the term is appropriate in the selected resources.

This section attempts to segregate the resources into certain categories but that is an artificial effort as many online resources are also available in hard copy.

Online Resources

The Military Education Home Page of the J-7 Operational Plans and Interoperability, the portion of the Joint Staff with responsibility for Joint Education, is available at http://www.dtic.mil/doctrine/. It gives detail to the doctrine, history, and training. It is a tremendous kick off for learning at this topic.

Another starting place for looking at resources on PME is the Air War College Web site entitled "Military Education Online," at http://www.a.af.mil/au/awc/awcgate/awc-pme.htm. This is an admirably comprehensive set of hyperlinks to other sites around the United States. The site is developed by the Air Force Institute for Advanced Distributed Learning, with periodic updates. The site gives not only a detailed description of the PME schools and their goals but includes indications whether the education is available through distance learning, nonresident courses.

One of the greatest advantages to pursuing information on PME is that the Services value their history and value discussions of the evolving state of their missions, activities, etc. As such, the vantage point of each Service is highlighted through Service histories, almost all available online. The Air Force service history is accessible at www.af.mil/history. The Army, with its Center for the Military History, has several Web sites such as www.army.mil/cmpg/ and broadly www.army.mil. West Point's history as the cradle of military education is explored at http://americanhistory.si.edu/westpoint/history_0.html, with various periods available through hypertext links. Navy issues appear at www.nwc.navy.mil as an entrance to the topic. Marine Corps topics open through www.marines.mil. Marine Corps University—Servicemembers' Online University accessed on January 22, 2006, at http://www.mcu.usmc.mil/sncoa/AAdegree.cfm.

The National Defense University has an entry into the online resources at www.ndu.edu. Its (NDU) library is accessible through the National Defense University Web site at http://www.ndu.edu/library. The reference material available through this site is remarkable, including the links to military journals at the Air University, the Joint Electronic Library, and various other online links. Additionally, the NDU Library also has links to various electronic databases (usually needing a password, however), and various other extraordinary connections in the PME scene.

One notable bibliography available through the NDU Library is the Goldwater-Nichols Bibliography at http://www.ndu.edu/library/pubs/Gol-Nich2.html#webref, which was accessible through a Google search. This bibliography contains articles, books, and other Web resources. Many of the articles have hyperlinks to other sites of relevance to PME topics.

The Air War College offers a link to a superb Navy War College library site at http://www.au.af.mil/au/awc/awcgate/mil-ed/nwc-bib.htm, which offers a bibliography containing many of the same items listed in this section.

Similarly, other PME institutions have a wealth of information as one would expect. The Combined Arms Research Library of the Command and General Staff College of the Army in Leavenworth, Kansas, is an example. The publications are detailed and written by practitioners such as retired General Paul F. Gorman's, *Secret of Future Victories* (Washington, DC: Institute for Defense Analyses, 1992), or other publications with details on what steps in PME have been taken by the United States through its history.

A fascinating account of President/General Dwight Eisenhower's role in the creation of the Industrial College of the Armed Forces and subsequent PME institutions is available at http://www.moaa.org/magazine/May2005/eisenhower.asp.

Articles

Breslin-Smith, Janet. "A School for Strategy: The Early Years." *Joint Force Quarterly*, 41(2) (2006): 40–44. This is an introductory essay on the history of the National War College, with a detailed discussion of the earliest years of the college.

DePuy, William E. "For the Joint Specialist: Five Steep Hills to Climb." *Parameters* (September 1989): 2–12. This is an assessment of the type of problems that the Joint Specialty Officers (JSOs) face in trying to achieve the necessary education and appropriate job selections to allow for promotion while also providing what is most important for national defense.

Gough, T. "Origins of the Army Industrial College." *Armed Forces and Society*, 17(2) (Winter 1991): 259–275. A discussion of the lessons the United States learned in World War I for providing logistics and mobilization in major conflict leading to the creation of a school for this purpose.

Greenwald, Byron. "The Anatomy of Change: Why Armies Succeed or Fail at Transformation." *The Land Warfare Papers* (35): 1–25.

Marné Peterson, Theresa. "America's National War College: Sixty Years of Educating Strategic Thinkers." *Joint Force Quarterly*, 41(2) (2006): 38–39. The Commandant of the National War College discusses the interrelationships and educational experience at the college.

Reyes Cordero, Miguel Ricardo. "International Colleagues at the National War College." *Joint Force Quarterly*, 41(2) (2006): 48–50. This essay outlines the role that international officers play in the educational experiences at the National Defense University.

Riker-Coleman, E. "The Case of David C. Jones." http://www.unc.edu/~chaos1/jones.pdf. An interesting essay on the man who first discussed and, probably unexpectedly, pushed for reform of the defense sector in the United States.

Sweeney, F.R. "The Army and Navy Staff College." *Command and General Staff School Military Review*, XXIII(4): 9–10.

Yaeger, John. "The Origins of Joint Professional Military Education." *Joint Force Quarterly*, 37(2nd quarter 2005): 74–82. Written by a Naval officer who served as Dean of the Industrial College where he is now director of institutional research. Yaeger discusses the beginning of PME at the Fort McNair campus and elsewhere in the early years of the twentieth century.

Yoshpe, H.B. "Bernard M. Baruch: Civilian Godfather of the Military m-day Plan." *Military Affairs*, XXIX(1): 15–30. A discussion of the man who drove mobilization and logistics in the United States for decades.

Monographs and Edited Volumes

Ambrose, Stephen. *Duty, Honor, Country: A History of West Point*. Baltimore, MD: Johns Hopkins University Press, 1999.

American Council on Education. *Handbook to the Guide to the Evaluation of Educational Experiences in the Armed Services*. Washington, DC: American Council on Education, 1995. The initial accreditation processes for PME institutions began with the council looking at these issues.

Anderson, Clinton and Steve Kim. *Adult Higher Education and the Military: Blending Traditional and Nontraditional Education*. Washington, DC: American Association of State Colleges and Universities, 1990. Increasingly fewer traditional academics have any exposure to how the military does anything. This is an introductory review of the types of pedagogies that the PME system utilizes.

Arnold, Edwin. *Professional Military Education: Its Historical Development and Challenges*. Carlisle Barracks, PA: U.S. Army War College, 1993. A basic tool to grasping the role and paths of PME in the United States.

Arnott, Gail. *Senior Service School Teaching Methods*. Maxwell Air Force Base, AL, 1989. Few outside the military have any understanding of the topic but Arnott introduced it as the PME accreditation process began after Goldwater-Nichols.

Ball, Harry. *Of Responsible Command: A History of the U.S. Army War College*. Carlisle Barracks, PA: Alumni Association of the U.S. Army War College, 1983. This is a study by an organization promoting the recognition of the Army War College in PME.

Bauer, T.W. *History of the Industrial College of the Armed Forces*. Washington, DC: Alumni Association of the Industrial College of the Armed Forces, 1983. Written as the Industrial College began its seventh decade as a school of "resourcing" national security strategy.

Betros, Lance, ed. *West Point: Two Centuries and Beyond*. Abilene, TX: McWhiney Foundation Press, 2004. Betros heads the History Department at the Academy and has pulled together a well-respected collection of essays on a range of questions to commemorate the end of the second century of West Point.

Biggs, Christopher. *Distance Education: A Case Study with Applications for DOD and the Marine Corps*. Monterey, CA: Naval Postgraduate School, 1994. In the 1990s as technology changed and budget reallocations were discussed, several PME institutions (along with civilian education) strongly considered the role of distance learning in their overall concept of education.

Boasso, Herbert. *Intelligence Support to Operations: The Role of Professional Military Education*. Maxwell Air Force Base, AL: Air University Press, 1988. The constant tension between how to include or exclude intelligence in education for military officers is a perennial topic for consideration in PME.

Bolinger, M. *Improving Officer Career and Intermediate Level Education*. Quantico, VA: Marine Corps CDC, 1991. While much discussion focuses on senior level PME, intermediate level actually affects more officers because it is almost universal.

Boston, James. *A History of the Role of Traditional Education for the United States Air Force Line Officer Corps, 1947–1995*. Fort Belvoir, VA: Defense Technical Information Center, 1996. Prior to the Goldwater-Nichols era, probably a higher percentage of Air Force officers pursued traditional master's degrees than any other service.

Broihier, Michael. *Applying Technology to Marine Corps Distance Learning*. Monterey, CA: Naval Postgraduate School, 1997. One of the difficulties distance learning has faced in the PME system, and traditional academic settings, has been trying to find its most useful applications based on technology available. This considered that carefully.

Brooks, Vincent K. *Knowledge Is the Key: Educating, Training, and Developing Operational Artists for the 21st Century*. Fort Leavenworth, KS: U.S. Army Command and General Staff College School of Advanced Military Studies, 1992. An Army general's study while a more junior officer at the School of Advanced Military Studies on what is needed for operational art, a term used most frequently by the Army, in the new century.

Burton, James G. *The Pentagon Wars: Reformers Challenge the Old Guard*. Annapolis, MD: Naval Institute Press, 1993. This is a discussion of the internal debate within layers of the Pentagon on the need and lack of need for reform, including professional military education.

Carlson, Adolf. *Advent of the Joint U.S. Army–Navy Warplanning on the Eve of the First World War: Origins and Legacy*. Carlisle Barracks, PA: Strategic Studies Institute, 1998. Few studies like this exist on the earliest joint planning but this explores the lessons and challenges.

Carver, George A. and Don M. Snider. *A New Military Strategy for the 1990s: Implications for Capabilities and Acquisitions*. Washington, DC: Center for Strategic and International

Studies, 1991. This is a discussion of the broad implications of the reform agenda at the beginning of the last decade. While some of the issues seem out-of-date, the commonalities of reform threatening some existing systems are important for professional military reform.

Clark, Edward. *A Comparative Analysis of Intermediate Service College (ISC) Phase I: Joint Professional Military Education (JPME)*. Monterey, CA: Naval Postgraduate School, 1990. As greater interest developed in the changes necessary to have PME support the goals of Goldwater-Nichols, studies of the two phases began to appear. This was an initial one.

Collins, John M. *Roles and Functions of the U.S. Combat Forces: Past, Present, and Prospects*. Washington, DC: Congressional Research Service, January 21, 1993. One of the "deans" of Washington analysts on defense issues, Collins writes thoughtfully of how role and functions would be affected by changes that education brings to the military hierarchy.

Conference of Professional Military Education for the 21st Century Warrior. *Proceedings of the Naval Postgraduate School and Office of Naval Research Conference on Military Education: PME for the 21st Century Warrior*. Monterey, CA: Naval Postgraduate School, 1998. See also http://web.nps.navy.mil/FutureWarrior/proceedings.html. This summary of the major PME conference at the Postgraduate School foretold many of the questions that have arisen about the topic over the past decade.

Crackel, T.J. *West Point: A Bicentennial History*. Lawrence, KS: University Press of Kansas, 2002. The bicentennial of Jefferson's founding of the Academy resulted in thorough reassessments such as this.

Crane, J. and J.F. Kieley. *United States Naval Academy: The First Hundred Years*. New York: McGraw-Hill Book Company, 1945. A century's education examined to commemorate the milestone.

———. *West Point: "The Key to America"*. New York: McGraw-Hill Book Company, 1947. A century and a half into the U.S. Military Academy, this volume considers the evolution of its educational practices and experience.

Crowe, William J. and David Channoff. *The Line of Fires: From Washington to the Gulf, The Politics and Battles of the New Military*. New York: Simon & Schuster, 1993. Admiral William Crowe, a plain-speaking admiral who served as the chairman of the Joint Chiefs of Staff as the 1986 military reform process was under debate, writes of the impacts that reform has had on the Services and the military education system. Holding not only the rank of four-star admiral but also a doctoral degree from Princeton in international relations, Crowe has a unique vision of professional military education because he rarely received education through that tract.

Donnini, Frank and Richard Davis. *Professional Military Education for Air Force Officers: Comments and Criticisms*. Maxwell Air Force Base, AL: Air University Press, 1991. Interesting set of observations about what would improve PME in the Air Force.

Ellis, J. and R. Moore. *School for Soldiers: West Point and the Profession of Arms*. New York: Oxford University Press, 1974. A scholarly treatment of what education means at West Point.

Flipper, Henry Ossian. *Black Frontiersman: The Memoirs of Henry O. Flipper, First Black Graduate of West Point*. Fort Worth, TX: Texas Christian University Press, 1997. A fascinating volume on what a former slave who went through the PME system that existed at the time experienced in his life, some good, some bad.

———. *The Colored Cadet at West Point*. New York: Lee, 1878; reprinted New York: Arno, 1898. Flipper was the first African-American to graduate from West Point and recounts his experience here.

Gebert, Stephen. *PME, Lessons Learned, and the Joint Operational Commander*. Newport, RI: U.S. Naval War College Joint Military Operations Department, 1998. This study considers what PME has contributed to one officer's view as a joint commander in the Navy.

Gilroy, Michael. *The End of History and the Last Soldier: Training Military Leaders to Operate with Information Superiority*. Newport, RI: U.S. Naval War College Joint Military Operations Department, 1998. In the late 1990s, much focus of the PME programs around the Services considered information warfare issues as paramount to studies.

Grigsby, Wayne. *The Current Interwar Years: Is the Army Moving in the Correct Direction?* Fort Leavenworth, KS: U.S. Army Command and General Staff College School of Advanced Military Studies, 1996. This is a cautionary study by a mid-level officer who is trying to see whether Goldwater-Nichols reforms were hurting or helping the Army a decade after the law went into effect.

Grum, Allen. *Use of Technologies in Education and Training: Army Science Board Ad Hoc Study Final Report*. Washington, DC: U.S. Army Science Board, 1995. One of the earliest considerations of how to integrate new technologies into training and education programs in the military.

Hammond, P.Y. *Organizing for Defense: The American Military Establishment in the Twentieth Century*. Princeton, NJ: Princeton University Press, 1961. A thorough examination of the various facets of putting together a security establishment, a generation after the 1947 reforms.

Hardesty, J. Michael. *Training for Peace: The U.S. Army's Post-Cold War Strategy*. Washington, DC: U.S. Institute of Peace, 1996. This is a study of the way that training and education now operate in the Army.

Hart, Gary and William S. Lind. *America Can Win: The Case for Military Reform*. Bethesda, MD: Adler & Adler, 1986. Former Colorado Democratic Senator Gary Hart long and passionately argued for defense reform much before it became seriously contemplated by many others. Hart's proposals included a grasp of the role of PME in this process.

Hattendorf, John, B. Mitchell Simpson, and J.R. Wadleigh. *Sailors and Scholars: The Centennial History of the U.S. Naval War College*. Newport, RI: Naval War College Press, 1984. A review of the contributions of this venerable curriculum and its students and faculty.

Hayes, J.D. and John Hattendorf. *The Writings of Stephen B. Luce*, Vol. 1. Newport, RI: Naval War College Historical Monograph Series, 1975. Foundational thoughts by the first president of the Naval War College.

Hollingsworth, Stephen. *The War Colleges: The Joint Alternative*. Maxwell Air Force Base, AL: Air War College, Air University, 1990.

Hunt, E. *History of Fort Leavenworth: 1827–1927*. Fort Leavenworth, KS: The General Service Schools Press, 1979. A history of the expansion of all aspects of this crucial Kansas facility, a majority of its efforts aimed at educating generations of Army officers.

Isenberg, David. *Missing the Point: Why the Reforms of the Joint Chiefs of Staff Won't Improve the U.S. Defense Policy*. Policy Analysis 100. Washington, DC: CATO Institute, 1988. Many critics rebutted General David Jones' and those views of others who supported reform. The CATO Institute, a libertarian think tank in the nation's capital, is another example of the opponents.

Jehart, Alojz. *Impact of the Revolution in Military Affairs on Education and Training of Professional Structures in Land Forces*. Washington, DC: Industrial College of the Armed Forces, 1997. An interesting analysis of how the Revolution in Military Affairs, the big theory of the 1990s, has affected land forces' education and its overall training concerns.

Jessup, P.C. *Elihu Root*. New York: Dodd, Mead, and Company, 1938. A biography of one of the most innovative organizers of the Army and defense establishment at the turn of the nineteenth century.

Karschinia, Paul. *Education, the War Colleges and Professional Military Development*. Washington, DC: The National War College, 1975. A view of the education process before the National Defense University and Goldwater-Nichols reforms began.

Kitfield, James. *Prodigal Soldiers: How the Generation of Officers Born of Vietnam Revolutionized the American Style of War*. New York: Simon & Schuster, 1995. Few analysts doubt that the officers who served in Vietnam, who then went on to leadership between 1985 and 2000 such as General Colin L. Powell and General Barry McCaffrey, made a tremendous impact on the thinking that now characterizes the uniformed services of the United States. Kitfield studies their careers.

Koczela, Diane and Dennis Walsh. *Promoting Distance Education at Naval Postgraduate School (NPS)*. Monterey, CA: Naval Postgraduate School, 1996. Many believed in the 1990s that distance learning answered the problems that the military had in getting people out of the field and into the classroom. This is an assessment along those lines.

Korb, Lawrence. *The System for Education of Military Officers in the U.S.* Pittsburgh, PA: International Studies Association, 1976. A scholarly appraisal of PME before the Goldwater-Nichols changes.

Lane, Randall. *Learning without Boundaries: The Future of Advanced Military Education*. Fort Belvoir, VA: Defense Technical Information Center, 1997. An analysis of the ways that military education can extend beyond its traditional boundaries.

Lederman, Gordon Nathaniel. *Reorganizing the Joint Chiefs of Staff: The Goldwater-Nichols Act of 1986*. Westport, CT: Greenwood Publishing, 1999. Written by an attorney more detached than the actors who fought the battle on this topic, it is a good summary history that gives perspective on how the Joint Staff process grew in the United States. The volume contains a fine bibliography divided into useful categories for people seeking to go further in this topic.

Leopold, R.W. *Elihu Root and the Conservative Tradition*. Boston, MA: Little, Brown, 1954. A further biography of the former secretary of war and state who changed PME in the Army.

Long, R. *National Defense University Transition Planning Committee Report*, NDU Collection 1989.

Lovelace, Douglas C. *Unification of the United States Armed Forces: Implementing the 1986 Department of Defense Reorganization Act*. Carlisle Barracks, PA: Strategic Studies Institute of the Army War College, 1996. Head of the prolific Strategic Studies Institute at the Army War College, Lovelace explains the importance, costs, and successes of bringing "jointness" to the U.S. defense establishment.

Lovell, John. *Neither Athens nor Sparta?: The American Service Academies in Transition*. Bloomington, IN: Indiana University Press, 1979.

Mahan, Alfred Thayer. *The Influence of Sea Power Upon History, 1660–1783*. Boston, MA: Little, Brown, 1893. One of the most renowned studies by any figure in U.S. military education, Mahan's lectures at the Naval War College were collected into a seminal volume that is still oft-quoted. It also marked the coming of age for the U.S. PME concept.

Masland, John and Lawrence Radway. *Soldiers and Scholars: Military Education and National Policy*. Princeton, NJ: Princeton University Press, 1957. An assessment of how national military education had affected the Services in the early years of the 1947 National Security Community.

Maurmann, S.F. *Air Force Instruction 36-2501: Officer Promotions and Selective Continuation*. Arlington, VA: U.S. Air Force, 2004. This report documents standards, including PME, for Air Force officers to succeed and rise through the ranks.

McNaugher, Thomas L. and Roger L. Sperry. *Improving Military Coordination: The Goldwater-Nichols Reorganization of the Department of Defense*. Washington, DC: The Brookings Institution, 1994. One of the major goals of PME is to facilitate "jointness" across the Services. Sperry and McNaugher discuss this as part of their study.

Miller, Stephen. *Joint Education: Where It Really Should Begin*. Carlisle Barracks, PA: U.S. Army War College, 1993. A provocative assessment of joint education.

Mitchell, Greig. *Application of Distance Learning Technology to Strategic Education*. Carlisle Barracks, PA: U.S. Army War College, 1996. Further discussion of distance learning as a solution to PME.

Moses, Louis J. *The Call for JCS Reform: Crucial Issues*. Washington, DC: National Defense University Press, 1985. Moses penned an assessment, as the Congress was moving toward reform in the mid-1980s, of what the needs were for reform.

Naval War College (U.S.). *History of the United States Naval War College, 1884–1963*. Newport, RI: U.S. Naval War College, 1964. A sanctioned history of the pride of Navy PME.

————. *Maritime Strategy Implementation: The Conceptual/Intellectual Infrastructure*. Newport, RI: U.S. Naval War College, 1987. The manifestation of Navy success on the cusp of Goldwater-Nichols reforms.

————. *Naval War College Strategic Plan: Mission Statement*. Newport, RI: U.S. Naval War College, 1995. The Naval War College shows its goals and how they will accomplish them.

————. *Report to the Leadership of the Navy: Past, Present, and Future*. Newport, RI: U.S. Naval War College, 1977. The Naval War College assesses its mission and its ability and challenges to meeting that mission.

Nenninger, Timothy. *The Leavenworth Schools and the Old Army: Education, Professionalism, and the Officer Corps of the United States Army, 1881–1918*. Westport, CT: Greenwood Press, 1978. A nicely worked discussion of the range of PME offered at the Army's venerable Fort Leavenworth for the last two decades of the nineteenth century through World War I.

Norton, Robert F. *The Impact of the Goldwater-Nichols Reorganization Act on the U.S. Army Reserve*. Carlisle Barracks, PA: U.S. Army War College, 1988. While most discussions focus on the active duty force, this study ponders the effects of reform on the Army Reserve component.

Odeen, Phillip, Andrew Goodpaster, and Melvin Laird. *Toward a More Effective Defense: The Final Report of the CSIS Defense Organization Project*. Washington, DC: Center for

Strategic and International Studies, Georgetown University, 1985. Odeen himself was quite prominent in the defense reorganization process and wrote an assessment, along with several other analysts, of the 1986 reforms.

Oskam, Margaretha I. *The Impact of the Goldwater-Nichols Reorganization Act of 1986 on the Woman Line Naval Officer.* Carlisle Barracks, PA: U.S. Army War College, 1988. Few studies separated out effects on women versus those on the entire force, particularly in the Navy. Oskam does this.

Palmer, D.R. *The River and the Rock: The history of Fortress West Point.* 2nd ed. West Point, New York: Hippocrene Books, 1991. This study discusses the physical reasons West Point was chosen for the national military academy and then begins on education there.

Pappas, G.S. *Prudens Furturi: The U.S. Army War College 1901–1967.* Carlisle Barracks, PA: The Alumni Association of the U.S. Army War College, 1967. This treatment describes the thinkers and strategies emanating from the Army War College.

Park, Soon-heon. *Evaluation of Graduate Education of the Military Professionals and Assessment of Their Needs.* Monterey, CA: Naval Postgraduate School, 1983. A pre-reform study of the aims and success of PME.

Parlier, Greg. *The Goldwater-Nichols Act of 1986: Resurgence in Defense Reform and Legacy.* Quantico, VA: Marine Corps Command and Staff College, 1989. Written with the goal of discussing how defense reform can go beyond the 1980s, this is a most interesting study when compared with calls for reform today, a generation later.

Partin, J.W. *A Brief History of Fort Leavenworth, 1827–1983.* Fort Leavenworth, KS: U.S. Army Command and General Staff College, 1983.

Pasierb, Barbara. *Educating the 21st Century Leader: A Critical View of the Military Senior Service Colleges with an Eye Toward Jointness.* Carlisle Barracks, PA: U.S. Army War College, 1998. This study questions whether the War Colleges are adequately meeting the jointness challenge.

Pentagon Library. *Defense Reorganization 1986–1990: A Selective Bibliography.* Washington, DC: Pentagon Library, 1990. A detailed listing of the range of materials on the first few years of Goldwater-Nichols reorganization actions.

Pickett, Dayton and Elizabeth Dial. *Joint Professional Military Education for Reserve Component: A Review of the Need for JPME for RC Officers Assigned to Joint Organizations.* McLean, VA: Logistics Management Institute, 1998. The attention given to preparation of any sort for the Reserve Component is always controversial because many feel it is not adequate.

Powers, James. *National Assistance and Civil–Military Operations: The Gap in Professional Military Education.* Carlisle Barracks, PA: U.S. Army War College, 1996. During a period of considerable military apprehension about civil–military issues, Powers assesses the problems that exist in current curricula.

Powers, Marcella. *A Survey of Studies Addressing Graduate Education on the United States Air Force.* Maxwell Air Force Base, AL: Air War College, Air University, 1987. A study of the quality of graduate education as PME itself is being restructured.

Preston, Richard. *Perspectives in the History of Military Education and Professionalism.* Colorado Springs, CO: U.S. Air Force Academy, 1980. An assessment of military professionalism and education as the Navy and Army were reevaluating their curricula after the Vietnam era.

Quinn, Dennis, ed. *The Goldwater-Nichols DOD Reorganization Act: A Ten-Year Retrospective*. Washington, DC: National Defense University Press, 1999. Accessible online at http://www.ndu.edu/inss/books/books%20-%201999/Goldwater-Nichols%20Retrospective%20-%20Nov%2099/GNDOD.pdf. Quinn edits a volume with half a dozen views on defense reorganization, written by three former senior military officers. He also gives views of how civilians operate in the upper echelons of the Defense organization.

Reading, P. *History of the Army and Navy Staff College*. Unpublished manuscript (copy held at National Defense University Library Special Collections, Washington, DC), 1972.

Rupinski, Timothy. *Selection Criteria for Professional Military Education*. Alexandria, VA: Center for Naval Analyses, 1987. One of the more interesting questions is how the Services allocate seats to their quota of students in each college. This study coincides with Goldwater-Nichols reform.

Shaw, Chris. *Professional Military Education: An Alternative Approach*. Washington, DC: Industrial College of the Armed Forces, 1992. A different interpretation of PME guidelines.

Shenk, Robert and Donald Ahern. *Literature in the Education of the Military Professional*. Colorado Springs, CO: English Department, U.S. Air Force Academy, 1982. The application of one discipline to military education.

Siegel, Adam. *A Brave New Curriculum for a Brave New World?* Alexandria, VA: Center for Naval Analyses, 1991. Many people asked whether the United States PME establishment was studying new challenges or old during the years immediately after the cold war, as does this author.

Simons, W.E., ed. *Professional Military Education in the United States: A Historical Dictionary*. Westport, CT: Greenwood Press, 2000.

Spector, Ronald. *"Professors of War": The Naval War College and the Modern American Navy*. New Haven, CT: Yale University Press, 1967. A premier military historian considers the role of the Newport education and the Navy.

Staten, Roddy and Lawrence Pemberton, Jr. *A Case Study of Distance Education and Its Application to the Marine Corps Institute (MCI)*. Monterey, CA: Naval Postgradaute School, 1995. This study of distance education looks at a specific case for application.

Sweatt, Owen. *Leadership as Teachership*. Carlisle Barracks, PA: U.S. Army War College, 1997. The Army, arguably more than the other Services, puts extraordinary emphasis on teaching leadership in PME at various levels.

Sweetman, J. *The U.S. Naval Academy: An Illustrated History*. 2nd ed. Annapolis, MD: Naval Institute Press, 1995. A history of the institution published by the Naval Institute.

U.S. Congress. House. Committee on Armed Services. Military Forces and Personnel Subcommittee. *Professional Military Education at the Armed Forces Staff College*. Hearing, 103rd Cong., 1st sess. Washington, DC: U.S. Government Printing Office, 1994. Homing in on the education experience for intermediate grade officers in Norfolk.

————. House. Committee on Armed Services. Panel on Military Education. *Advanced Military Studies Programs at the Command and Staff Colleges*. Hearings, 102nd Cong., 2d sess. Washington, DC: U.S. Government Printing Office, 1993. All the Services have intermediate colleges which this hearing considered.

————. House. Committee on Armed Services. Panel on Military Education. *Oversight Hearings*. Hearings, 101st Cong., 1–2d sess. Washington, DC: U.S. Government

Printing Office, 1991. The testimony of Admiral Kurth and Captain Wylie are recommended.

————. House. Committee on Armed Services. Panel on Military Education. *Professional Military Education*. Hearings, 100th Cong., 1–2 sess. Washington, DC: U.S. Government Printing Office, 1990.

————. House. Committee on Armed Services. Panel on Military Education. *Professional Military Education*. Hearings, 102d Cong., 1st sess. Washington, DC: U.S. Government Printing Office, 1992.

————. House. Committee on Armed Services. Panel on Military Education. *Report of the Panel on Military Education of the One Hundredth Congress*. 101st Cong., 1st sess. Washington, DC: U.S. Government Printing Office, 1989.

————. *Report of the Panel on Military Education of the One Hundredth Congress*. Committee Print, 101st Cong., 1st sess. Washington, DC: U.S. Government Printing Office, 1989.

U.S. Department of Defense. Committee on Excellence in Education. *The Senior Service Colleges: Conclusions and Initiatives*. Washington, DC: Department of Defense, 1975.

————. Office of the Inspector General. *Joint Professional Military Education: Inspection Report*. Arlington, VA: Office of the Inspector General, 1993.

————. *Joint Professional Military Education: Phase II, JPME Evaluation Report*. Arlington, VA: Office of the Inspector General, 1998.

U.S. General Accounting Office. *Report to the Air Force: Status of Recommendations on Officers' Professional Military Education: Briefing Chairman, Panel on Military Education, Committee on Armed Services, House of Representatives*. Washington, DC: U.S. Government Printing Office, 1991.

————. *Army: Status of Recommendations on Officers' Professional Military Education: Briefing Report to the Chairman, Panel on Military Education, Committee on Armed Services, House of Representatives*. Washington, DC: U.S. Government Printing Office, 1991.

————. *Department of Defense: Professional Military Education at the Intermediate Service Schools: Report to the Chairman, Panel on Military Education, Committee on Armed Services, House of Representatives*. Washington, DC: U.S. Government Printing Office, 1991.

————. *Department of Defense: Professional Military Education at the Senior Service Schools: Report to the Chairman, Panel on Military Education, Committee on Armed Services, House of Representatives*. Washington, DC: U.S. Government Printing Office, 1991.

————. *DOD Service Academies: Academy Preparatory Schools Need a Clearer Mission and Better Oversight: Report to the Chairman, Senate and House Committees on Armed Services*. Washington, DC: U.S. Government Printing Office, 1992.

————. *Marine Corps: Status of Recommendations on Officers' Professional Military Education: Fact Sheet for the Chairman, Panel on Military Education, Committee on Armed Services, House of Representatives*. Washington, DC: U.S. Government Printing Office, 1991.

————. *Military Education: Actions on Recommendations Involving Institute for National Strategic Studies and Capstone: Report to the Chairman, Panel on Military Education, Committee on Armed Services, House of Representatives*. Washington, DC: U.S. Government Printing Office, 1992.

————. *Military Education: Curriculum Changes at the Armed Forces Staff College: Report to the Chairman, Panel on Military Education, Committee on Armed Services, House of Representatives*. Washington, DC: U.S. Government Printing Office, 1991.

————. *Military Education: Implementation of Recommendations at the Industrial College of the Armed Forces: Report to the Chairman, Panel on Military Education, Committee on Armed Services, House of Representatives.* Washington, DC: U.S. Government Printing Office, 1992.

————. *Military Education: Implementation of Recommendations at the National War College: Report to the Chairman, Panel on Military Education, Committee on Armed Services, House of Representatives.* Washington, DC: U.S. Government Printing Office, 1992.

————. *Military Education: Implementation of Recommendations at the Armed Forces Staff College: Report to the Chairman, Panel on Military Education, Committee on Armed Services, House of Representatives.* Washington, DC: U.S. Government Printing Office, 1991.

————. *Military Education: Information on Service Academies and Schools: Briefing Report to Congressional Requesters.* Washington, DC: U.S. Government Printing Office, 1993.

————. *Military Training: Cost-Effective Development of Simulations Presents Significant Challenges: Report to Congressional Committees.* Washington, DC: U.S. Government Printing Office, 1995.

————. *Military Training: Potential to Use Lessons Learned to Avoid Past Mistakes Is Largely Untapped: Report to the Chairman, Subcommittee on Military Personnel, Committee on National Security, House of Representatives.* Washington, DC: U.S. Government Printing Office, 1995.

————. *Navy: Status of Recommendations on Officers' Professional Military Education: Briefing Report to the Chairman, Panel on Military Education, Committee on Armed Services, House of Representatives.* Washington, DC: U.S. Government Printing Office, 1991.

————. *Professional Military Education: Testimony.* Washington, DC: U.S. Government Printing Office, 1991.

————. *Service Academies: Historical Proportion of New Officers During Benchmark Periods: Report to Congressional Requesters.* Washington, DC: U.S. Government Printing Office, 1992.

U.S. Office of the Chairman of the Joint Chiefs of Staff. *Military Education Policy Document.* Washington, DC: U.S. Government Printing Office, 1993.

Van Creveld, Martin. *The Training of Officers: From Military Professionalism to Irrelevance.* New York: Free Press, 1990. Van Creveld is a scholar advocating a reconsideration of the role of education in the preparation of officer corps.

Walker, James. *International Military Education and Training: The Ultimate Foreign Policy Tool for the 21st Century.* Carlisle Barracks, PA: U.S. Army War College, 1998. Walker considers the role of international officers education in the creation of similar programs for the United States.

Walsh, Daniel. *Joint Professional Military Education and Its Effects on the Unrestricted Line Naval Officer Career.* Monterey, CA: Naval Postgraduate School, 1997. Each of the Services had its own reaction to the impact of changes in PME. This considers Navy effects.

Weiss, Michael. *The Education and Development of Strategic Planners in the Navy.* Monterey, CA: Naval Postgraduate School, 1990. With the advent of PME reform affecting all Services, the question posed by this volume is crucial.

Wesley, E.B. *Proposed: The University of the United States.* Minneapolis, MN: The University of Minnesota Press, 1936. In the early decades of the last century, the concept of a national university for civil servants and uniformed officers was popular.

Whitley, A.D. *Armed Forces Staff College: Command History, 1946–1981.* Norfolk, VA: National Defense University Press, 1981. The current Joint Forces Staff College has a much longer history, discussed here, as the AFSC.

Widen, S. Craig. *United States Military Cultures: A Mandatory Lesson for Senior College Curriculum.* Carlisle Barracks, PA: U.S. Army War College, 1997. This study contemplates the role that military culture plays in PME at various senior service schools.

Wilsbach, Kenneth. *United States Air Force Operational Education, Training and Organization.* Newport, RI: U.S. Naval War College, Joint Military Operations Department, 1998. This study is highly aimed at the operation of the Air Force while conducted at a sister Senior Service institution.

Wilson, James. *Postgraduate Education and Professional Military Development: Are They Compatible?* Monterey, CA: Naval Postgraduate School, 1991. This study asks the question that many people fear asking: Does PME in any way undermine military ethos or operations?

Winkler, John and Paul Steinberg. *Restructuring Military Education and Training: Lessons from Rand Research.* Santa Monica, CA: Rand, 1997. This study, a decade after Goldwater-Nichols went into effect, notes that there are lessons available from reforms with possible applications for the future.

Think Tank Papers and Studies

Carafano, James Jay and Alane Kochems. "Rethinking Professional Military Education." *Executive Memorandum #976.* Washington, DC: The Heritage Foundation, July 28, 2005, accessed on January 22, 2006 at http://www.heritage.org/Research/NationalSecurity/em976.cfm.

Goodpaster, Andrew, L. Elliott, and J. Allen Hovey. *Towards a Consensus on Military Service: Report of the Atlantic Council's Working Group on Military Service.* New York: Pergamon Press, 1982. A study of military education is embedded in this report, one of the authors a former commandant of the National War College.

Science Applications International Corporation. *Conference Report: Professional Military Education and the Emerging Revolution in Military Affairs, 22–23 May 1995 National Defense University.* Rockville, MD: SAIC, 1995. In the 1990s, the PME community wrestled with how much to take from the Revolution in Military Affairs (RMA).

Taylor, Bill, ed. *Professional Military Education: An Asset for Peace and Progress.* Washington, DC: Center for Strategic and International Studies, 1997. This study, chaired by Richard Cheney, examined the role of PME in the U.S. force. It is thorough and comprehensive in its assessment of jointness in the field a decade after major reforms, which included military education.

U.S. General Accounting Office. *Defense Personnel: Actions Planned to Implement Reorganization Act.* AGO/NSIAD-88-157 (April). Washington, DC: Government Printing Office, 1988. The General Accounting Office (now Government Accountability Office) examined the concerns about Goldwater-Nichols, creating adverse effects that could downgrade recruiting and retention.

———. *Defense Personnel: Status of Implementing Joint Assignments for Military Leaders.* AO/NSIAD-91050BR (January). Washington, DC: General Accounting Office, 1991.

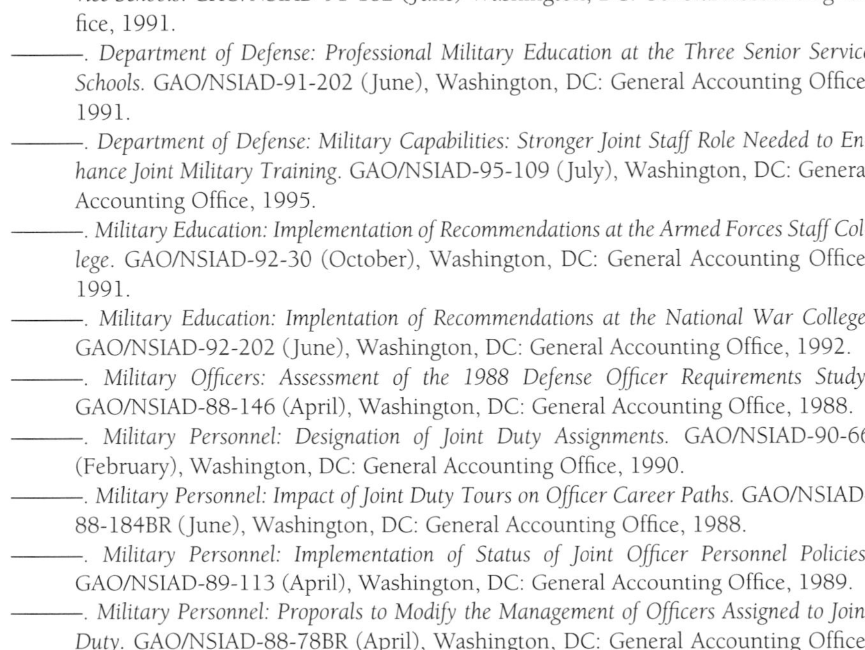

———. *Department of Defense: Professional Military Education at the Four Intermediate Service Schools.* GAO/NSIAD-91-182 (June) Washington, DC: General Accounting Office, 1991.

———. *Department of Defense: Professional Military Education at the Three Senior Service Schools.* GAO/NSIAD-91-202 (June), Washington, DC: General Accounting Office, 1991.

———. *Department of Defense: Military Capabilities: Stronger Joint Staff Role Needed to Enhance Joint Military Training.* GAO/NSIAD-95-109 (July), Washington, DC: General Accounting Office, 1995.

———. *Military Education: Implementation of Recommendations at the Armed Forces Staff College.* GAO/NSIAD-92-30 (October), Washington, DC: General Accounting Office, 1991.

———. *Military Education: Implentation of Recommendations at the National War College.* GAO/NSIAD-92-202 (June), Washington, DC: General Accounting Office, 1992.

———. *Military Officers: Assessment of the 1988 Defense Officer Requirements Study.* GAO/NSIAD-88-146 (April), Washington, DC: General Accounting Office, 1988.

———. *Military Personnel: Designation of Joint Duty Assignments.* GAO/NSIAD-90-66 (February), Washington, DC: General Accounting Office, 1990.

———. *Military Personnel: Impact of Joint Duty Tours on Officer Career Paths.* GAO/NSIAD-88-184BR (June), Washington, DC: General Accounting Office, 1988.

———. *Military Personnel: Implementation of Status of Joint Officer Personnel Policies.* GAO/NSIAD-89-113 (April), Washington, DC: General Accounting Office, 1989.

———. *Military Personnel: Proporals to Modify the Management of Officers Assigned to Joint Duty.* GAO/NSIAD-88-78BR (April), Washington, DC: General Accounting Office, 1988.

Winkler, John D. and Paul Steinberg. *Restructuring Military Education and Training: Lessons from Rand Research.* Santa Monica, CA: RAND, 1997. The RAND Institution, with its long ties to research and analysis of the existing programs of various portions of the U.S. Government, produced a solid study of what would be most useful in PME at the turn of the century, a decade after *Operation Desert Storm* and the end of the cold war.

Journals on Professional Military Education

These journals concentrate almost exclusively on military issues, with a significant component in military education concerns. They are readily accessible in libraries.

Army Times, published by *Army Times*, accessible at www.armytimes.com

Armed Forces Journal, published by *Army Times*, accessible at www.armedforcesjournal.com

Joint Force Quarterly, published by the National Defense University Press, accessible at www.ndu.edu/inss

Marine Corps Times, published by *Marine Corps Times*, accessible at www.marinecorpstimes.com

Naval War College Review, published by the Naval War College

Navy Times, a commercial paper, published by *Navy Times*, accessible at www.navytimes.com

Parameters, published by the Army War College

Theses and Dissertations

A'Hearn, F.W. "The Industrial College of the Armed Forces: Contextual Analysis of an Evolving Mission, 1924–1994." Unpublished dissertation. Blacksburg, VA: Virginia Polytechnic Institute and State University, 1997.

Boggs, Kevin G., Dale A. Bourque, Kathleen M. Grabowski, Harold K. James, and Julie K. Stanley. "The Goldwater-Nichols Department of Defense Reorganization Act of 1986: An Analysis of Air Force Implementation of Title IV and Its Impact on the Air Force Officer Corps." Unpublished thesis. Maxwell Air Force Base, AL: Air Command and Staff College, 1995. The first decade of Goldwater-Nichols had many officers wondering whether the joint requirements would affect promotions. This is a study of whether the first eight years had borne out that fear.

Gatliff, Robert E. and Mary C. Pruitt. "Title IV of the Department of Defense Reorganization Act of 1986: Hidden Impacts." Unpublished thesis. Maxwell Air Force Base, AL: Air War College, 1988. As this title indicates, there were a number of unintended consequences from the Goldwater-Nichols reforms. This examines how crucial they are.

Kovach, John Peter. "An Analysis of Naval Officers Serving on Joint Duty: The Impact of the 1986 Goldwater-Nichols Act." Unpublished thesis. Monterey, CA: Naval Postgraduate School, 1996.

Savage, Dennis. "Joint Duty Prerequistites for Promotion to Genreal/Flag Officer." Unpublished thesis. Carlisle Barracks, PA: U.S. Army War College, 1992.

van Trees Medlock, Kathleen. "A Critical Analysis of the Impact of the Defense Reorganization Act on American Officership." Unpublished dissertation. Fairfax, VA: George Mason University, 1993.

Yaeger, John W. "Congressional Influence on National Defense University." Unpublished dissertation. Washington, DC: George Washington University, 2005. A tremendous piece of work on how congressional reforms have affected the joint PME at NDU over the past five decades.

Young, Terry J. "Title IV—Joint Officer Personnel Policy: A Peace Dividend Is Required." Unpublished thesis. Carlisle Barracks, PA: U.S. Army War College, 1992.

U.S. Government Publications

Bibliography of Professional Military Education (PME), compiled by Greta E. Marlatt. Monterey, CA: Naval Postgraduate School, January 1998: 21 pp. Available online at http://web.nps.navy.mil/~library/bibs/pmetoc.htm.

Study on Military Professionalism (1970). A brutally frank discussion of problems that the Army had created in prosecuting the war in Vietnam such as the weekly "body count" statistics instead of meaningful measurements of success or failure in reaching the end state desired.

United States. Chairman of the Joint Chiefs of Staff. *Officer Professional Military Education Policy (OPMEP)*. December 31, 2005. Washington, DC: JCS, 2005. Reissued periodically, the *Officer Professional Military Education Policy* is the Chairman of the Joint Chief of Staff's guidance to PME institutions on what is necessary for officers to study to achieve the level of certification that will grant officers the required educational status under the 1986 Goldwater-Nichols Military Reform Act. Once officers have achieved Phase I and Phase II accreditation, they are eligible to assume billets that have been designated as "joint," hence meeting requirements that are necessary for officers to reach "general" or "flag" (in the Navy) rank.

United States. Congress. House. Committee on Armed Services. Panel on Military Education. *Advanced Military Studies Programs at the Command and Staff Colleges.* Hearing before the Panel on Military Education of the Committee on Armed Services, House of Representatives, 102nd Cong., 2d sess.: hearing held May 12, and July 23, 1992. Washington, DC: U.S. Government Printing Office, 1992 [CIS 93-H201-25]. The House of Representatives, particularly Congressman Ike Skelton of Missouri, has had a long, passionate interest in military education and periodically asks the military to explain its changes in military education.

————. Congress. House. Committee on Armed Services. Panel on Military Education. *Executive Summary.* Committee Print. Washington, DC: U.S. Government Printing Office, 1989. This is the Committee hearing executive summary for the year at the end of the decade when Goldwater-Nichols went into effect and Congressman Ike Skelton, the Missouri democrat, suggested significant reforms to the PME system in place.

U.S. Institute of Higher Defense Studies. *Capstone, Syllabus, General and Flag Officer Professional Military Education Courses, Joint and Combined Studies.* Washington, DC: National Defense University, 1985. Capstone was the next to last major piece of the PME system in the United States, aimed at newly promoted flag and general officers who have not had the exposure to combined and joint strategic concerns. This is the syllabus and basic information for a program required under Goldwater-Nichols.

United States. Joint Chiefs of Staff. *A Strategic Vision for the Professional Military Education of Officers in the Twenty-first Century.* Washington, DC: JCS, 1995. This is somewhat broader and explanatory than the Officer Professional Military Education Plan issued periodically by the chairman of the Joint Chiefs Office.

Professional Military Reading Lists

Each of the following institutions or offices offer a reading list of professional military material that it suggests for professional development. The lists are available on the National Defense University Web page which is accessible at http://www.ndu.edu/info/ReadingList.cfm.

Chairman of the Joint Chief of Staff's Military Reading List
Chief of Staff of the Air Force Military Reading List
The Coast Guard Commandant's Reading List
Congressman Ike Skelton's Military Article List
Congressman Ike Skelton's Military Reading List
Industrial College of the Armed Forces Reading List
Joint Forces Staff College Commandant's Reading List
Navy Professional Reading List
NDU President's Military Reading List
The U.S. Army Chief of Staff's Reading List

About the Author

CYNTHIA A. WATSON has been Professor of Strategy at the National War College since 1992, serving as a core course director, Associate Dean of Faculty and Curriculum, Director of Faculty Development, and Director of Electives. She earned a master's degree from the London School of Economics and a doctorate from the University of Notre Dame. Dr. Watson has authored several books on security issues, including *Nation-Building* (2004), *U.S. National Security* (one of *CHOICE's* top books of 2002), *U.S. National Security Interest Groups* (Greenwood, 1990), and was contributing coeditor of *Political Role of the Military* (Greenwood, 1996). She has written many articles as well. Her current research interests revolve around China's growing role in Latin America.